Energy Trading & Hedging

ENERGY TRADING & HEDGING

A NONTECHNICAL GUIDE

TOM SENG

> **Disclaimer**
> The recommendations, advice, descriptions, and methods in this book are presented solely for educational purposes. The author and publisher assume no liability whatsoever for any loss or damage that results from the use of any of the material in this book. Use of the material in this book is solely at the risk of the user.

Copyright © 2019 by
Endeavor Business Media, LLC
1421 South Sheridan Road
Tulsa, Oklahoma 74112-6600 USA

800.752.9764
+1.918.831.9421
sales@pennwell.com
www.pennwellbooks.com

Publisher: Matthew Dresher

Cover image: © Getty images - FroYo_92

Library of Congress Cataloging-in-Publication Data Available on Request

Names: Seng, Thomas, 1955- author.
Title: Energy Trading & Hedging / Thomas Seng.
Description: Tulsa, Oklahoma, USA : PennWell Corporation, [2019] | Includes bibliographical references.
Identifiers: LCCN 2019011132 (print) | LCCN 2019021869 (ebook) | ISBN 1593706375 (ebook) | ISBN 1593704747 (hardcover)
Subjects: LCSH: Petroleum industry and trade. | Futures market.
Classification: LCC HG6047.P47 (ebook) | LCC HG6047.P47 S45 2019 (print) | DDC 332.64/52—dc23
LC record available at https://nam04.safelinks.protection.outlook.com/?url=https%3A%2F%2Flccn.loc.gov%2F2019011132&data=02%7C01%7Cmattd%40pennwell.com%7C7ecc24ddec6b4f82e84508d6ce3f7d26%7C5bbf75da8a3f493c8343e6cd0cb0e070%7C0%7C0%7C636923169417327419&sdata=cIt%2Br044v5Ft0Sx5KFwYRP3Ldz06nDTtxEJkCmqm6tI%3D&reserved=0

All rights reserved. No part of this book may be reproduced, stored in a retrieval system, or transcribed in any form or by any means, electronic or mechanical, including photocopying and recording, without the prior written permission of the publisher.

Printed in the United States of America

1 2 3 4 5   22 21 20 19

This book is dedicated to my wife,
Mary and our three children, Bailey, Michael and, Evan.
They are the reason I get up every day.

Contents

Acknowledgments xi
Introduction xiii
    Notes xix

Chapter 1
Supply/Demand Fundamentals for Energy Prices 1
    Key Learning Points 1
    Crude Oil 2
        Weather 2
        US economy 4
        International economy 5
        Currency 6
        Geopolitical events 8
        Supply and demand statistics 11
        Cross-commodity markets 12
    Natural Gas 13
        Weather 13
        US economy 14
        Supply and demand statistics 14
        Fuel switching 16
    Review Exercises 19
    Notes 21

Chapter 2
Oil and Natural Gas Cash Markets 23
    Key Learning Points 23
    Natural Gas and Crude Oil—Physical Pricing 25
    The Intercontinental Exchange (ICE) Trading Platform 27
        Column labels 28
    Summary Points 30
    Review Exercises 31
    Notes 32

Chapter 3
Financial Energy Commodity Markets. 33
    Key Learning Points 33
    Energy Futures Contracts 35
    Summary Points 39
    Review Exercises 40
    Notes 41

Chapter 4
The New York Mercantile Exchange 43
    Key Learning Points 43
    History of the New York Mercantile Exchange 44
    NYMEX Contract Specifications 45
    Settlement Procedures 51
    Summary Points 52
    Review Exercises 53
    Notes 54

Chapter 5
Mechanics of Futures Markets 55
    Key Learning Points 55
    NYMEX Futures Quotes 56
        Column labels. 56
    Types of Orders 57
    Strips 58
    Margins 59
    Summary Points 61
    Review Exercises 62
    Notes 63

Chapter 6
Using NYMEX Contracts for Trading and Hedging. 65
    Key Learning Points: NYMEX Trading. 65
    Trading of NYMEX Futures Contracts 65
    The Commitments of Traders Reports:
        A Measure of Speculation and Hedging 67
    Risk Management and Hedging Using NYMEX Futures Contracts
        Key Learning Points: Energy Risk Hedging 67
    Hedging. 68
    Summary Points 76
    Review Exercises 77
    Notes 78

Chapter 7
Financial Energy Derivatives: Swaps 79
    Key Learning Points: Financial Energy Derivatives—Swaps. 79
    Basis Swaps. 83
    Using Basis Swaps to Hedge Transportation 89
    Summary Points 91
    Review Exercises 92
    Notes 93

Chapter 8
Financial Energy Derivatives: Spreads. 95
    Key Learning Points: Spreads 95
    Spreads 95
    The Hedging of Storage Capacity Using Time Spreads 98
    Summary Points 100
    Review Exercises 101
    Notes 102

Chapter 9
Financial Energy Derivatives: Options 103
    Key Learning Points: Options Contracts 103
    Options 104
    Options Models. 106
    Characteristics of Options. 107
    Hedging Using Options 107
    Summary Points 110
    Review Exercises 111
    Notes 112

Chapter 10
Technical Analysis. 113
    Overview. 113
    Key Learning Points 113
    Trend Lines. 117
        Elements of a technical chart 119
        Price signals 121
    Summary Points: Technical Analysis 125
    Review Questions 127
    Notes 128

Chapter 11
Risk Controls in Energy Commodity Trading and Hedging 129
    Overview. 129
    Key Learning Points: Risk Controls in Energy Commodity Trading
        and Hedging. 129
    Case Study 1: Barings Bank, PLC. 130
    Case Study 2: Orange County, CA. 131
    Case Study 3: Metallgesellschaft AG (MG) 132
    Key Lessons Learned by Examining the Case Studies. 133
    Risk Policies and Controls for Energy Commodity Derivatives. 133
    Summary Points 137
    Review Exercises 138
    Notes 139

Appendix A
EIA's "Weekly Petroleum Status Report: Highlights"
for Week Ending January 25, 2019 141
    Note 142

Appendix B
EIA's "Weekly Natural Gas Storage Report" for Week Ending January 25, 2019....... 143
    Note 144

Appendix C
Energy markets risk management Glossary 145

Acknowledgments

The contents of this book are largely based upon the "on-the-job" training that I and my friends and colleagues received over the years in the energy industry. I organized the topics for teaching purposes and supplemented my knowledge by researching other books on these subjects as well as, various online data sources. It is also the result of teaching as an Adjunct university Instructor for (14) years and, as a full-time Assistant Professor for the past (5) years. The practical examples I provide are literally how business is conducted on the marketing and trading floors across the US. So, in a general way, I want to thank all of those I learned with along the way during my time in the industry. More specifically, I have two "brothers-in-arms" in trading, Tim and Scott, who have become lifelong friends. I need to acknowledge Ted Jacobs, the former Director of Energy Management at both *The University of Oklahoma* and, *The University of Tulsa*, who got me started teaching energy marketing at the university level. That forced me to organize the various aspects of marketing & trading I had experience firsthand into lectures and to continually shape my courses based upon this ever-changing industry. I want to thank Erin Long, with *Penn State University's* John A. Dutton e-Education Institute, for helping me to organize my first-ever online course there. That experience allowed me to shape the content of this book. I also owe thanks to Lisa Ponti and the *Global Association of Risk Professionals* (GARP) for allowing me to use their extensive glossary of risk terminology. I also want to acknowledge *The University of Tulsa* for giving me the opportunity to teach both graduate and, undergraduate courses in energy which furthered my understanding of the topics within this book. Lastly, I want to thank Steve Hill, Matt Dresher and the rest of the PennWell publishing company personnel I have dealt with in getting this book in print.

Introduction

No practitioner can survive in the modern commodity markets without a solid understanding of derivatives.

—Vincent Kaminski

As the saying goes, "The only constant is change." This statement can certainly be used to describe the energy industry over the past several decades. Booms and busts have occurred numerous times as prices have risen and fallen, and thousands of companies have come and gone. Enron shook the very foundation of energy trading with its abuse of mark-to-market accounting, leading to the eventual collapse of the company. In addition, in the early 2000s, investigations were conducted concerning supply and price manipulation in both the gas and power sectors. The investigations resulted in fines and imprisonment for many energy traders, contributing to the disappearance of the former top-five energy marketing companies in the United States.

The banking industry rushed in to provide the lost financial counterparty liquidity. A fledgling electronic trading platform, the Intercontinental Exchange (ICE),[1] replaced Enron Online as the dominant over-the-counter electronic marketplace. In 2008, however, the banks encountered their own financial problems, and to a large extent, they also exited the energy commodity sector.

New exploration, drilling, and completion techniques, such 3-D and 4-D seismic, horizontal and directional drilling, and hydraulic fracturing (also called *fracking*) emerged on the energy scene. These advances not only led to a substantial increase in the US production of crude oil and natural gas, but also to great controversy over the methods used and the new regulations enacted. Nonetheless, these sources of energy supply have changed the nature of global energy markets. The United States became a rising force in the international energy markets as an exporter of crude oil and liquefied natural gas (LNG), along with refined products, natural gas liquids (NGLs), and petrochemical feedstocks, such as ethylene and propylene. (Late in 2015, the first shipments of crude oil exports left the United States for the first time since a 1975 ban. In February 2016, the first shipment of LNG left the Louisiana Gulf Coast and thus began a new era for the domestic US natural gas industry.)

Additionally, the United States stands on the verge of energy self-sufficiency,[2] which could dramatically change its policies toward intervention in oil-rich areas of the world, once considered areas of "strategic interest." (While the United States could potentially produce all of the oil and gas it needs, so long as it is engaged in the global markets for these commodities, domestic prices will be influenced by those dynamics. In addition, a substantial increase in US crude oil refining infrastructure is needed to reduce the volume of refined products imported.)

Price volatility for these commodities has increased dramatically over the past several years. In July 2007, the price per barrel (bbl) of US domestic crude oil was approximately $70. One year later, in July 2008, the price reached a historic high of $147/bbl. Then, just five months after that, prices fell to around $35/bbl. Figure I–1 indicates the month-ending price for July 2008, and figure I–2 indicates the month-ending price for December 2008. Both are different from the previously quoted daily prices.

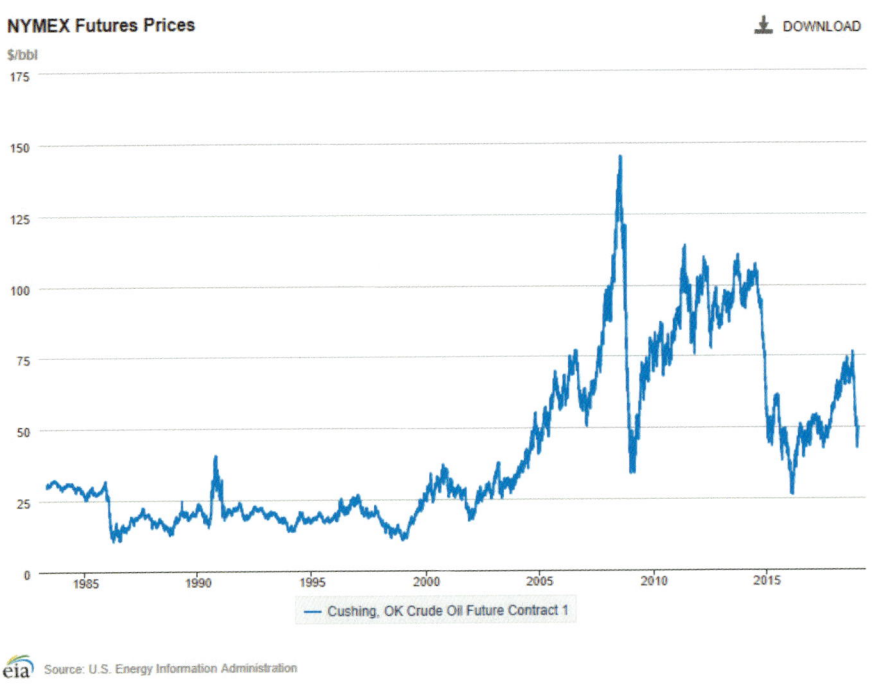

Fig. I–1. Cushing crude oil futures contract for July 2008

A host of factors contributed to this dramatic change in price, but the initial reaction of the market was to blame "those traders in New York," a reference to the "speculative" traders in the pits of the New York Mercantile Exchange (NYMEX). In fact, more traders had entered the market in the form of hedge funds, private equity investors, and others seeking a safe harbor from the volatility of the stock market,

Introduction xv

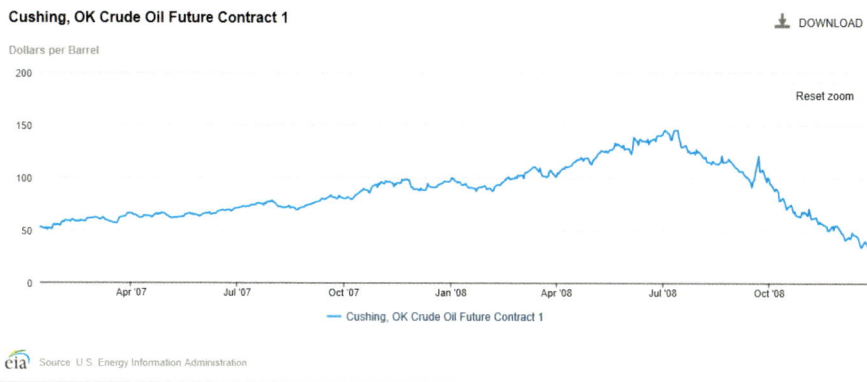

Fig. I–2. Cushing crude oil futures contract for December 2008

the US dollar, and other financial investments. In addition, it was revealed that many of those investing in crude oil contracts were international institutions and individuals. As a result, crude oil gained recognition as a truly global commodity, with a host of new factors influencing price.

US crude oil reached $100 USD/bbl in June 2014 (fig. I–3), only to fall below $50 USD/bbl six months later as Saudi Arabia declared economic war on the

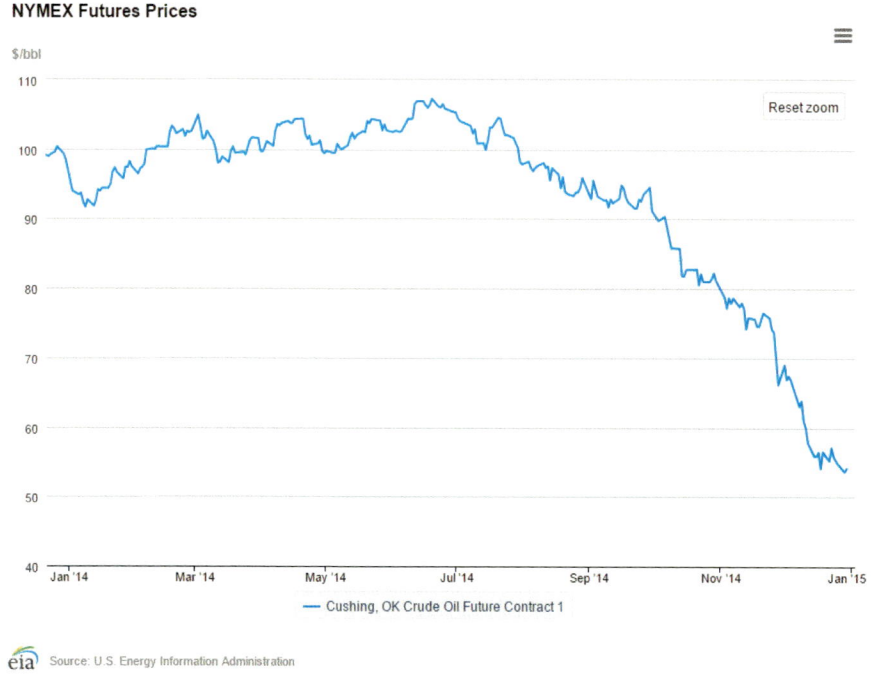

Fig. I–3. Cushing crude oil futures contract for June 2014

US shale oil producers (fig. I–4). Fearing a loss of market share to US oil and gas companies, the Saudis "opened the valves" and produced as much oil as they could. By August 2015, prices were $45 USD/bbl and fell to about $26/bbl in February 2016. The oil industry in the United States would face hardship the likes of which had not been seen since the bust years of the early 1980s.

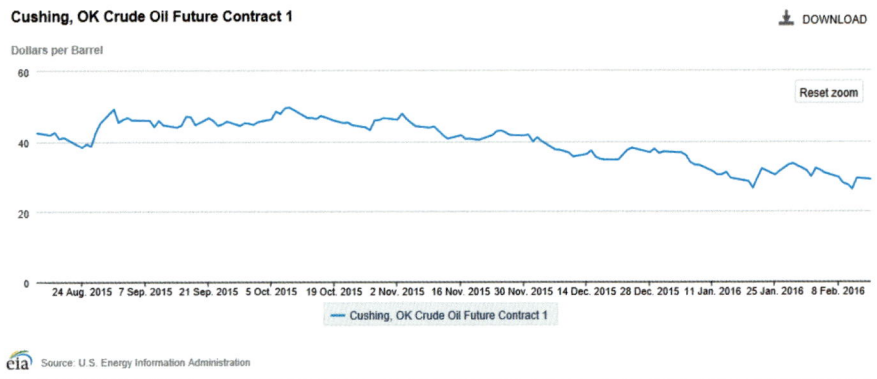

Fig. I–4. Cushing crude oil futures contract for February 2015

This text will focus primarily on the energy products that are financially traded on the New York Mercantile Exchange: crude oil, natural gas, unleaded gasoline,

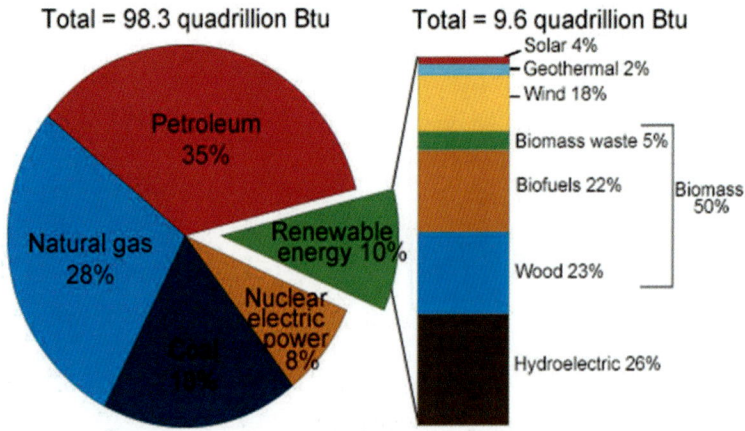

Fig. I-5. US energy consumption by energy source, 2014

and heating oil. These commodities are the most common ones used for hedging energy commodity risk. Each of these products has a profound effect on the US and international economies. Billions of dollars of infrastructure and millions of jobs worldwide are involved in exploration, production, processing, refining, transportation, storage, marketing, and distribution of these forms of energy. And economies cannot run without them. In fact, recent data by the Energy Information Administration, part of the Department of Energy, shows that only 11% of the energy consumed in the US comes from *all* sources of "alternative/renewable" energy including, solar, wind and, hydroelectric. (fig. I-5) Furthermore, as illustrated in fig. I-6, there was only a 1% growth in this area of energy production between 2014 and 2017. According to the EIA's latest "Annual Energy Outlook 2019", only 31% of the country's electrical generation in 2050 will come from these sources.[1] Petroleum and natural gas usage will remain strong for decades to come.

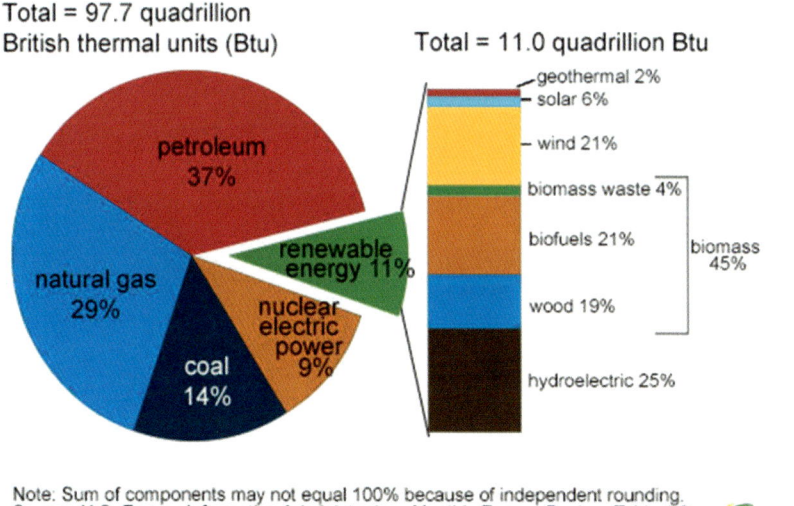

Fig. I-6. US energy consumption by energy source, 2017

Price is a key consideration when entering into any business venture. Potential revenue cannot be estimated without knowing what price the market will bear for the goods and services produced. Fortunately for energy commodities, there is an active financial market that allows for price transparency and the opportunity to reduce both price and physical risk.

---

1. https://www.eia.gov/todayinenergy/detail.php?id=38112

In this book, we will discuss several of the factors that can influence energy price direction, the financial instruments used in the market, and how commercial entities utilize these instruments to hedge the risks inherent in energy production and consumption, i.e., price and physical risk. We will also address the use of energy financial derivatives strictly for profit (*speculative trading*), and the exchanges that make efficient and competitive trading possible. The physical cash market will be addressed, along with the major industry publications that produce market prices, known as *postings* or *indexes*. Additionally, we will briefly discuss *technical analysis*, the use of charts to identify price trends and predict directional changes. Finally, we will discuss the need for stringent controls on the trading of financial energy derivatives to avoid some of the more notorious pitfalls that have led to the collapse of major global companies and institutions.

Notes

1. The successful online brokerage ICE, based in Atlanta, Georgia, eventually bought the International Petroleum Exchange (IPE) in London, renaming it ICE Futures Europe. Later acquisitions included the NYSE.
2. There is a semantic argument over the use of the terms *energy independence* and *energy self-sufficiency*. The former implies no interaction with the international markets that would impact US oil and gas prices, while the latter indicates enough energy production to meet the US demand.

1

Supply/Demand Fundamentals for Energy Prices

Key Learning Points

- There are several factors that can influence the price direction of energy commodity markets.
- The extent to which a particular factor actually impacts price direction has more to do with the *perception* of its potential impact by market participants than whether or not there is a direct impact.
- Crude oil is a globally traded commodity, and the United States has surpassed Saudi Arabia and Russia as the world's top oil producer, due in part to the shale oil "revolution" in the United States.
- The increase in natural gas supply coming from the shale plays has led to a surplus, resulting in LNG exports as a new market outlet.
- Weather is a key factor in determining demand for energy, but it impacts natural gas more than crude oil.
- The northeastern region of the United States is the world's largest consumer of heating oil due to the continued use of boilers for hot water and radiant heat.
- The use of coal for electrical power generation has decreased as the use of natural gas for power generation has increased, especially during the peak summer air-conditioning season.
- A future increase or decrease in demand for energy can be inferred from US and global economic data. Daily, weekly, monthly, quarterly, and annual economic data, from sources both private and governmental, are researched and analyzed by market participants.
- Geopolitical conditions and events can impact oil prices if regions of production are involved.
- Actual physical changes in commodity production, consumption, and inventories are reflected in the supply/demand data.
- Alternative and renewable energy sources are growing steadily but will take decades to represent a substantial share of the US energy mix.

- Traders focus on two key inventory reports released weekly by the US Energy Information Administration, one for crude and one for natural gas.

There are a host of factors that can potentially influence the direction of energy prices, some more obvious than others. It is important for anyone engaged in the financial energy markets to be aware of these and to research them on a continual basis. Whether one is trading for profit or watching the markets for an opportunity to hedge, an understanding of the information that impacts energy prices is extremely valuable. (It should be noted that although gathering supply and demand data can aid in making a better-informed market decision, it does not guarantee success.)

A recurring theme throughout this book is consideration of the human factor. We need to keep in mind that market participants are humans[1] and may react to market news in an emotional way. After all, they are in the market to make money, and those who act quickly stand to make the most profit. Conversely, market particpants do not wish to lose money, either.

So the results of market players' actions set the prices in financial energy markets. (When we address technical analysis in a future chapter, it will become apparent that the charts are really representations of the results of human interaction in the marketplace. We then attempt to determine the probability that people will react in the same way at some future point in time. That can offer clues to price direction.)

Below are several fundamental factors that might influence energy prices for crude and natural gas in the United States. This is certainly not an all-inclusive list. Many factors, such as weather and driving mileage, are obvious, but others are not. Again, keep in mind that these are factors that could influence energy prices to the extent that those in the marketplace perceive them to be important.

Crude Oil

The US standard for crude oil is West Texas Intermediate (WTI), a light, "sweet" crude.[2] The key market hub in the United States for crude is the Cushing Interchange, a pipeline and storage hub in Cushing, OK. Some of the factors that can actually impact supply and demand, or merely imply a change in supply and demand, are discussed in the following.

Weather

Heating oil is a refined distillate of crude oil. According to the EIA, "Heating oil ranks as the third most important source of residential energy in the Nation, with about 7.5 percent of all households using heating oil as their primary space heating fuel."[3] It is widely used today in the Northeast United States as a fuel in boilers to create hot water, radiant heat, and to a much lesser extent, to generate electricity. In fact, the northeastern part of the United States is the world's largest consumer of heating oil. As a result, cold weather in this region indicates the potential for increased demand for heating oil (i.e., crude oil). Since the other regions of the United States do not rely on heating oil to the same extent, if at all, cold weather in those areas is not viewed as a factor that influences the price of oil.

In addition to weather data such as temperature and windchill, heating degree days (HDDs) are a good indicator of winter demand. *Heating degree days* represent the theoretical amount of energy needed to raise an office building's temperature to a baseline 65°F when the outside temperature is lower than the temperature inside. (Historical and forecasted HDDs can be found on the website for the National Oceanic and Atmospheric Administration.[4])

Thus if a city, state, or region's HDDs are above normal, it means more energy was used for heating purposes. The converse is also true, as less energy consumption is indicated by fewer HDDs than normal. (*Note:* HDDs can never be negative. At temperatures above 65°F, *cooling degree days* [CDDs] are measured, as it then becomes a question of how much energy it takes to *lower* the building's temperature to 65°F.)

Another weather event the energy market keeps a watchful eye out for is the development of hurricanes, which could impact the US Gulf of Mexico's oil and gas drilling and production platforms. According to the US Energy Information Administration, 17% of the country's offshore oil production and more than 45% of all US oil refining capacity is located in and around the Gulf of Mexico.[5] At a minimum, if a tropical storm enters the gulf, operators evacuate offshore platforms about 24 hours in advance of the storm. Once landfall occurs, platforms are remanned, with the entire process taking at least 72 hours. This shuts in a large portion of the oil and gas production, and estimates of these outages are normally provided by the US Bureau of Ocean Energy Management (BOEM), formerly the US Minerals Management Service (MMS).

At the other end of the spectrum, an intense hurricane could develop with the potential to damage or destroy platforms and rigs. In terms of offshore damage, in 2005 Hurricane Katrina damaged 12 rigs and 30 platforms, with 18 of those platforms completely destroyed. In 2004, Hurricane Ivan damaged 7 rigs, destroyed 2 rigs and 7 platforms, and damaged 100 underwater pipelines.[6]

One of the ironies of hurricanes in the gulf is that once they reach landfall, they bring rain and cooler temperatures, which lower energy demand immediately. The swath of a tropical system can be wide and far-reaching, impacting US weather for days.

The official hurricane season begins on June 1 each year and ends on November 30, with the peak occurring around September 11.[7] During this time, the Weather Channel © provides a "Tropical Update" at 50 minutes past every hour, a critical report for those watching energy markets.

US economy

Energy runs the economy, so economic factors of all kinds are indications of energy consumption or the prospect of consumption. Given that crude oil is a global commodity, both US and international economic conditions can impact prices.

In the United States, the most obvious and most frequently reported indication of economic health is the stock market. The Dow Jones Industrial Index, the S&P 500, and NASDAQ results offer a daily assessment of the future health of the US economy from the perspective of investors, both domestic and foreign. Gains in the stock market imply future economic growth, which translates into an implied increase in future energy use.

In addition, there are weekly, monthly, and quarterly economic reports that can have an immediate impact on price perception, as well. These include the following (not an exhaustive list):

1. US Department of Labor, Bureau of Labor Statistics, unemployment reports (https://www.bls.gov/news.release/empsit.nr0.htm /)—Thursdays.
2. Institute for Supply Management Index[8] (http://www.ism.ws/index.cfm)—monthly. (ISM is a nonprofit organization that reports on activity in the manufacturing sector.)
3. "Consumer Price Index," US Bureau of Labor Statistics (http://www.bls.gov/cpi/)—monthly. The bureau publishes various indices related to consumer prices as measures of inflation. Lower inflation gives consumers, who represent 50% of the economy, more "disposable" income, which can spur manufacturing and service industry activities. The impact on energy demand will not be immediately apparent and could actually take months to occur. But the expectation is that the goods and services purchased will result in increased economic activity.
4. US gross domestic product (GDP), US Bureau of Economic Analysis (http://bea.gov/newsreleases/national/GDP/GDPnewsrelease.htm)—quarterly. The GDP is a key measure of a nation's economic health and standard of living. A recession is considered to have occurred after two consecutive quarters of declining GDP.
5. The US Department of Commerce's Economics and Statistics Administration (ESA) has a vast number of economic indicators, including data from the US Census Bureau and the US Bureau of Economic Analysis. These include areas such as the following:
   - Construction spending
   - Housing starts
   - Housing sales, both new and existing
   - US international trade (import vs. export deficit)
   - Monthly wholesale trade
   - Manufacturing and trade
   - Sales for retail and food services

- Personal income
- Personal spending

6. Annual and quarterly earnings reports from US companies.

International economy

In as much as we now recognize crude oil as a globally traded commodity, world economic conditions can influence prices in a similar fashion as that of the US domestic economy. Certain countries and regions can have a greater influence on oil pricing than others. As such, their economic activity becomes a critical barometer of crude oil demand.

When trading starts each morning on the exchanges in New York, Asian markets have already closed, and European markets are at midday or early afternoon. Thus signs of how markets here will potentially trade are already known as the early trends tend to continue, barring any news to the contrary. These include equities, currencies, and energy futures markets.

Perhaps the most-watched national economy these days, outside of the United States, is China's. As of the end of 2013, China's GDP represented approximately 15% of the world's economy, having grown about 500% since 2005 alone.[9] (See fig. 1–1.)

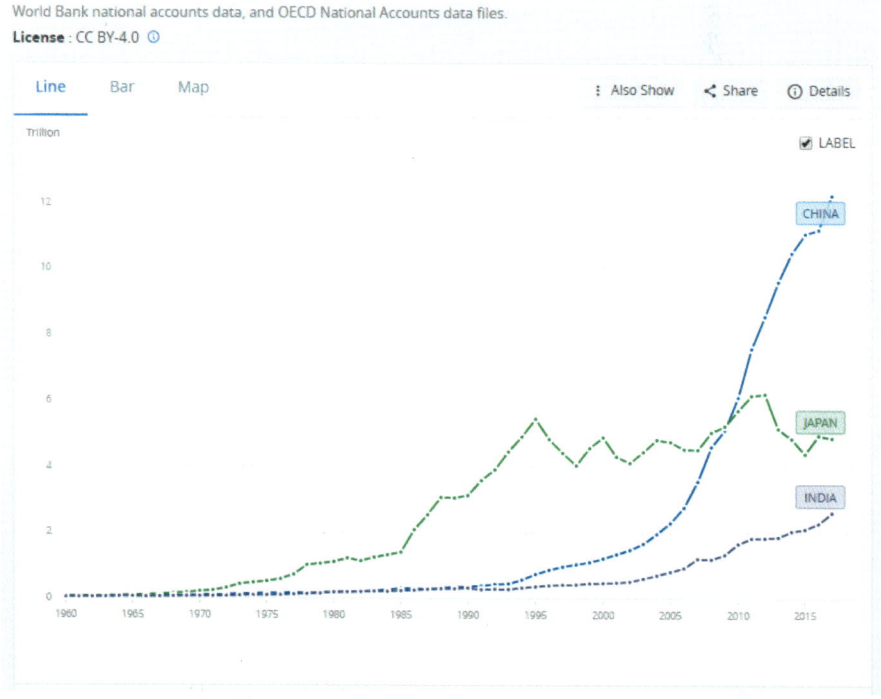

Fig. 1–1. China/Japan/India GDP growth 1960 - 2017

China became the world's largest energy consumer in 2011.[10] According to the US Energy Information Agency, China surpassed the United States as the world's largest net importer of oil and petroluem liquids in September 2013.[11] Using the same mindset as regards the US economy, oil prices can fluctuate with the changes in China's economy.

Coming in behind China in crude oil consumption is Japan, whose Fukushima nuclear disaster in 2011 shifted its energy mix to a higher degree of fossil fuels, making them the second-largest importer of all fossil fuels in 2013.[12] Japan's GDP has also risen steadily since 2008, as seen in figure 1–1.

Another country with an emerging economy in 2008 was India. Along with China, India was given partial credit for the run in crude oil prices due to increases in manufacturing and exports. The country's GDP took a big leap in 2008 and continued higher before stabilizing in 2012 (fig. 1–1). In 2015, India ranked fourth in the world in crude oil and products consumption, behind only the United States, China, and Japan.[13] It is expected that India will move into the second spot in oil consumption within a few years, surpassing even China.

In 2013, Europe imported about 22% of the world's crude oil.[14] Trade in Europe is largely regulated by the European Union, which is currently comprised of 26 continental European countries, plus Ireland and Great Britain. (Great Britain has voted to leave the EU and is expected to officially depart in 2019.) The EU has struggled economically in recent years, due in large part to the financial problems of Portugal, Ireland, Greece, and Spain (not so politely referred to as the "PIGS"), with Greece seeming to teeter on bankruptcy somewhat frequently. The International Monetary Fund (IMF) and the European Central Bank (ECB) have struggled to keep Greece afloat. When bankruptcy looms for any of these countries, or as bailout plans are proposed, oil prices ebb and flow. The global economy these days is so intrinsically tied to various national economies that the domino effect of a nation in default could be far-reaching.

The United Kingdom, formerly a member of the European Union and its largest producer of oil, became a net importer of oil in 2013, largely due to declining North Sea production. Oil represented 38% of all energy consumed in the United Kingdom in 2016.[15]

Currency

Among the new traders in crude oil futures in 2008 were international investors, who also had to consider currency valuations in their market involvment. Crude oil on the global market is traded in US dollars, so fluctuations in the value of the US dollar against other currencies can influence oil price direction. As the value of the US dollar fluctuates against other currencies, buying power shifts for international investors. For example, an investor in the United Kingdom can buy more crude oil futures contracts than a US investor when the British pound sterling (GBP) has a higher value than the US dollar (USD), as explained in the example below. This influx of foreign capital tends to raise the price of crude oil. Conversely, during periods of a strong USD, foreign investors tend to sell off oil futures contracts and invest in other instruments. This can lead to lower prices. In fact, while it does not occur 100% of the time, there is a strong inverse correlation between the price of crude and the USD. Figure 1–2 illustrates the relationship between crude oil and the value of the USD when compared to the GBP. In fact, if this chart were folded down the middle, the two halves would nearly be mirror images.

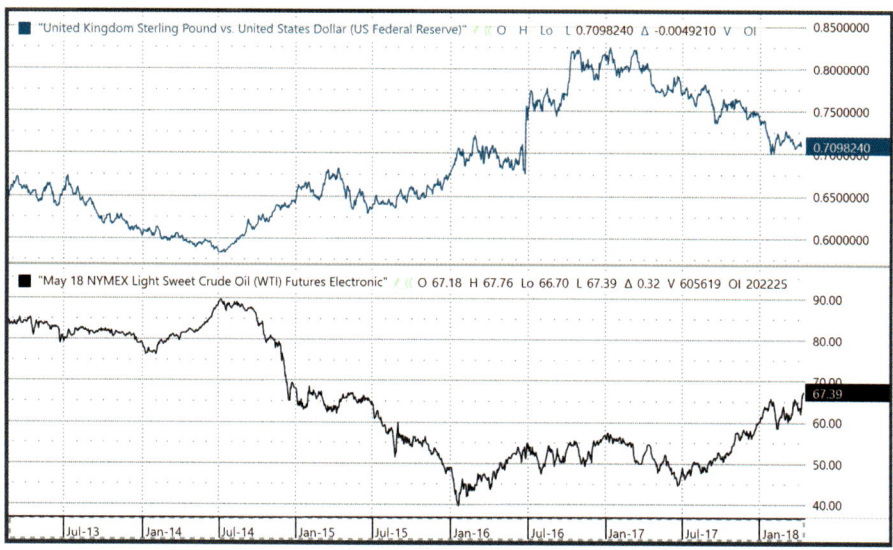

Fig. 1–2. Valuation comparison between US dollar, British pound sterling, and WTI pricing

Example: Foreign currency buying power. This concept can be illustrated as it relates to crude oil purchases:
1. US crude oil = $50 USD/bbl; GBP = $1.50 USD.
2. US investor can purchase 1,000 bbl for $50,000 USD.
3. UK investor can purchase 1,500 bbl for £50,000 ($75,000 USD).

It is readily apparent that tracking a currency marker such as the *Wall Street Journal's* USD Index can help traders predict movements in crude prices. While currency movements relative to one another can appear to be miniscule, keep in mind that billions and billions of dollars change hands daily. (In early 2018, China announced plans to launch a crude oil exchange that would be based upon their currency, the yuan. This could certainly impact the correlation mentioned here.)

Geopolitical events

Perhaps the most far-reaching impact on global energy prices is that caused by geopolitical conflicts, which often have an immediate and dramatic effect on prices. Again, there does not have to be an actual supply disruption, just the *perception* of one, for traders to react in the marketplace. Once the "smoke clears" and the facts are known, the market tends to settle down and return to its previous price trend.

Unfortunately, world history illustrates that there will always be countries battling over something, whether it is land or natural resources. (In his now-famous book, *The Prize: The Epic Quest for Oil, Money & Power*, Daniel Yergin makes a strong case for oil being the true underlying reason behind most major global conflicts over the past century.) But each situation has to be weighed on its ability to interrupt the production and delivery of crude oil out of that region. As of this writing, there are several conflicts watched by oil markets.

1. *Israel/Gaza (Hamas)*. Despite the fact that Israel is a minor producer of oil and refined products, this ongoing territorial battle emphasizes continuing tensions in the Middle East, an area of significant oil production.
2. *Nigeria—*. MEND, or the Movement for the Emancipation of the Niger Delta, and NDA, the Niger Delta Avengers, are both rebel groups in Nigeria. Such groups have repeatedly destroyed oil infrastructures, including pipelines and loading terminals belonging to international oil companies (IOCs) and the state-run oil company, Nigerian National Petroleum Corporation (NNPC).
3. *Russia/Ukraine*. Russia is now the world's 2nd-largest producer of oil behind the US.[16]
4. *ISIS*. ISIS, or the Islamic State of Iraq and Syria, is a radical Muslim terrorist group that invaded parts of both Iraq and Syria, overtaking towns and slaughtering their inhabitants. These religious extremists controlled producing oil fields and refineries in northern Iraq for nearly two years, raising concerns about supplies of crude and refined products in the region. ISIS sold crude on the black market to finance the caliphate's terrorist activities. While the ISIS hold on northern Iraq and Syria has been nearly eradicated, nothing but total annihilation will end their reign of terror.
5. *South Sudan*. In 2011, a referendum in Sudan led to the establishment of this separate country, which now controls the majority of oil production and reserves from the formerly unified region. But civil unrest, mostly along tribal lines, has interrupted production and brought uncertainty to the country's future.
6. *Iran*. The country's ongoing nuclear program, while said to be for energy production only, has been suspected of developing enriched

uranium for nuclear weapons. As a result, sanctions imposed against the regime resulted in a halt in the production and sale of oil. In January 2016, an agreement was reached which lifted the sanctions, allowing Iran to restart its oil production and begin exports of crude again. As of early 2017, Iran was producing about 3.7 million bbl/d, moving toward their goal of 4.0 million bbl/d, their presanction production level. Further tensions between Iran and the West center around Iran's supposed support of pro-Assad forces in Syria. Additionally, the United States has withdrawn from the agreement and reenacted sanctions against Iran.

7. *Oil politics.* In mid-June 2014, WTI was trading at about $107/bbl. By the end of the year, it was down to about half that price at $54/bbl. Two main factors were blamed for the plummeting prices. First, the increased supplies of oil in North America coming from sources once considered nonconventional, such as shale and tight formations, were outpacing demand. Second, Saudi Arabia, fearful of losing its position as the world's leading oil supplier[14] due to the increasing North American oil production, flooded the market with cheap crude in an effort to maintain its market share. This had a severe economic impact on countries dependent on oil revenue, such as Russia, Nigeria, and Venezuela, among others. Even Saudi Arabia eventually felt the impact and cut popular subsidies for its people. In the United States, many E&P companies cut back on exploration and drilling or halted activity altogether, resulting in hundreds of thousands of lost jobs. But in late 2016, OPEC and several noncartel members (including Russia) voted to voluntarily curtail production output. The lower oil prices had hurt the economies of most of the OPEC member nations and Russia. Effective in January 2017, the voluntary cuts moved global oil prices back into the range of $50/bbl to $55/bbl, spurring new activity in the United States. The higher prices tested the US O&G industry's ability to quickly respond to the gap in global oil production. Rig rates increased, and production climbed back to 9.0 million bbl/d in early 2017. It became economical to bring online wells that were previously "drilled but uncompleted" (DUC). By early 2018, OPEC and the cooperating noncartel countries agreed to extend the cuts to at least the end of that year, pushing the price per barrel steadily into the $60s for WTI and $70s for Brent.

8. *US politics.* There has always been, and probably will always be, conflict between the states that produce energy and those who largely consume it. These battles play out on a regular basis in Congress, at the White House, and even sometimes in front of the US Supreme Court. Some of the key issues in the past few years have included the following:

a. Hydrofracturing. This controversial well completion technique is being opposed or prohibited outright in many states. In addition, the high-pressure disposal of the wastewater into underground reservoirs is being linked to an increase in seismic activity in some states.[18]

b. Keystone XL Pipeline. Transcanada's proposed pipeline project, which would bring more bitumen[19] to the Cushing, OK energy hub, was denied US presidential approval for the US/Canada border crossing, part of Phase I of the project, in 2016. Phase II of the project, shipping crude from Cushing to the eastern Texas refining corridor, has already been completed and is operational.

c. Dakota Access Pipeline (DAPL). Energy Transfer Company's crude oil pipeline project, designed to bring North Dakota Bakken crude to Midwest refineries and pipeline hubs, faced protests from Native American tribes concerned about possible contamination of their water supply should a leak occur, although the pipeline route was on adjacent and not native lands.

d. Oil export ban of 1975. President Gerald Ford signed into law the "Energy Policy and Conservation Act" in December 1975, in response to the Arab oil embargo of 1973. It gave the President the authority to ban export of all petroleum products and established the Strategic Petroleum Reserve. Ronald Reagan later lifted the ban on export of refined products[20] such as condensate, a light oil produced at lease sites or through natural gas processing. In December 2015, Congress lifted the export ban, and crude oil was once again exported from the United States. By late 2018, the United States was exporting approximately 3.0 million bbl/d of crude oil.[21]

e. The "ruling party." US President Barack Obama had a more proenvironmental stance, as illustrated by such agreements as the Paris Climate Agreement and various new Environmental Protection Agency (EPA) restrictions on greenhouse gases (GHGs), including the Clean Power Plan. Fossil fuel usage was discouraged during his time in office. In January 2017, Donald Trump succeeded Obama as President and had a Republican Senate and House of Representatives to work with. He was very supportive of oil and gas, even going so far as to appoint Rex Tillerson, CEO of Exxon, to Secretary of State. He later appointed Oklahoma's attorney general, a litigant of the EPA, to chair that agency. He also signed executive orders in support of the controversial Keystone XL and Dakota Access oil pipelines, and the EPA rolled back many regulations deemed too restrictive on the oil and gas industry.

These are but a few of the many geopolitical situations that can, and do, influence crude prices.

Supply and demand statistics

Any information concerning oil production and/or consumption of its distillate products is basic supply/demand information that is central to the market. Numerous sources publish this data on a regular basis, including governmental entities (US and international) and industry associations and groups.

Some select sources include the following:
- British Petroleum's annual *Statistical Review of World Energy*[22]
- International Energy Agency (IEA)
- US Energy Information Administration (EIA)
  - *This Week in Petroleum* – provides weekly production, imports/exports and, refinery utilization.
- American Petroleum Institute (API)—weekly crude inventory report
- American Automobile Association (AAA)—miles driven
- Baker Hughes' rig count[23]
- US EIA's "Weekly Petroleum Status Report."[24] Each week, the EIA issues a report on its website of crude oil and its various distillates. The report is issued at 10:30 AM (Eastern US time) on Wednesdays and encompasses all reported activity as of the prior Friday. Several pieces of key supply and demand statistics are included in this report. Energy commodity traders are focused on the release of this information, as it can have an immediate impact on price direction. (See appendix A for a copy of the report.)

Key components of the report include the following:
1. *Refinery utilization.* The percentage of total US refinery capacity that is running, which indicates demand for crude, as well as production of gasoline and other distillates. Since crude is not used in its raw form, refineries represent the only source of demand for oil.
2. *Imports.* Both raw crude oil and refined products such as gasoline are imported, and volumes are compared to "year-ago" data, indicating an improving or worsening balance. With the large increase in domestic oil production, the amount of imported crude and petroleum products has been steadily declining, as shown in figure 1–3.
3. *Commercial crude oil inventories.*[25] The change in US inventory from one week to the next has a profound impact on oil prices from a pure trading standpoint. Analysts provide forecasts for the change in inventory ahead of the actual report. Financial energy commodity traders react to the

difference between forecasted and actual inventories. This reaction occurs instantaneously, as savvy traders are glued to the EIA website awaiting the report, often hitting the "refresh" button in hopes of getting the information before their competitors.

4. *Gasoline inventories.* Total gasoline products, as well as a breakdown between finished gasoline and blending products, gives a picture of supply and demand for gasoline. A decrease in total products could mean more demand for refinery feedstocks. A surplus could mean just the opposite. To further interpret this data, it needs to be understood that *blending components* are merely the 10%–15% of additives necessary to arrive at the 90/10 and 85/15 blends of unleaded gasoline at the retail pump (also called *finished gasoline*).

5. *Distillate fuel.* This category is primarily comprised of heating oil. As indicated above, cold weather translates into increased use of heating oil in the northeastern United States. Thus, the amount of distillate fuel in storage during the winter months becomes a critical factor.

Cross-commodity markets

The price of Brent North Sea crude oil, the global standard, can influence the price of WTI as the two tend to move in tandem. Brent is traded on the ICE Futures–Europe exchange. The lifting of the US ban on exports in December 2015 allowed WTI to compete in the global markets, thus closing the gap between the US and global standards. Additionally, crude distillates prices for products such as gasoline, diesel, heating oil, and jet fuel directly impact the price of crude.

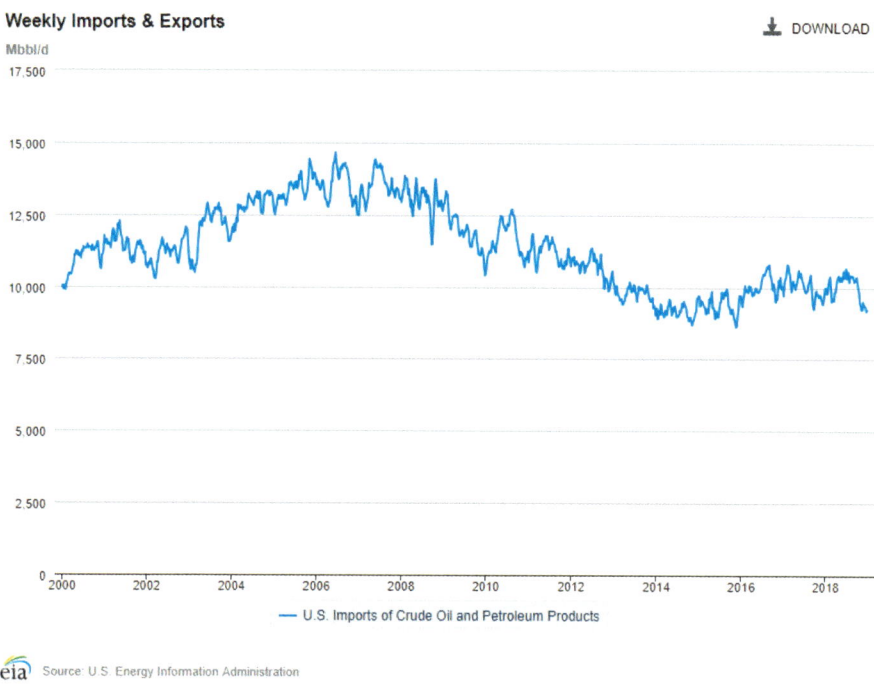

Fig. 1–3. US imports of crude oil and petroleum products, declining since 2010

Natural Gas

Unlike WTI, the North American natural gas market is driven more by domestic factors than global ones and is more seasonally impacted.[26] As such, not all of the factors that influence crude oil prices apply to natural gas prices. For instance, geopolitical events and global economic conditions currently have little impact on natural gas price direction. The following are some of the more relevant factors influencing North American natural gas prices.

Weather

Natural gas is used for both residential hot water and space heating, as well as to generate electricity. So both winter and summer seasons can have an impact on natural gas prices, depending on the region. Extreme winters such as that of 2013/2014 can lead to supply shortages and record-high prices, as can other weather events, such as "bomb cyclones"[27] and polar vortexes. Excessively hot summers can also cause price spikes for natural gas, especially in the southern, southwestern, and western United States. Power plants in these regions often generate gas-fired electricity to satisfy air-conditioning loads. The converse is also true in that mild winters and summers can result in surplus inventories and lower prices. As with oil, HDDs are an important indicator of demand for natural gas during cold weather, but cooling degree days (CDDs) must be monitored during the hotter periods. Additionally, as mentioned in the case of crude, hurricanes in the Gulf of Mexico also can disrupt natural gas supplies.

Another weather event, unique to the US Pacific Northwest, is that of "snow pack". The NW US is the largest area in the country for hydroelectric power generation. But, that generation is dependent upon the amount of snowfall that occurs each winter that will, then, result in the Springtime "run-off" to feed the rivers downstream where the hydroelectric dams are located. So, the natural gas market watches snowfall totals throughout this region of the country.

US economy

Since North American natural gas is not yet being exported as LNG in large quantities,[28] global economic conditions are not viewed as having a *direct* impact on prices. However, the many potential domestic economic factors that can influence crude oil, listed above, have to be considered for natural gas as well.

Supply and demand statistics

As with crude, the same sources of data on production and consumption for natural gas provide basic fundamental information that can translate into price signals. The EIA website is the best source for up-to-date and historical data. As the United States exports increasing amounts of natural gas to Mexico in the form of LNG, those volumes will have a direct impact on the supply/demand balance in this country.

US EIA's "Weekly Natural Gas Storage Report." One important source of information is the US EIA's "Weekly Natural Gas Storage Report." In the same fashion that the EIA reports changes in crude oil and distillates inventory, they also publish a weekly report on the status of US natural gas inventories. Each Thursday at 10:30 AM (US Eastern time), the EIA releases its "Weekly Natural Gas Storage Report,"[29] representing the prior week's change in natural gas inventory at the country's underground storage facilities. (See appendix B for a sample copy of the EIA's report for the week ending January 25, 2019.)

The report shows the following:

1. *Regional breakdown.* The activity is tracked for each of the EIA-defined regions (fig. 1–4), which are designat.ed as the East, Midwest, Mountain, Pacific, and South Central regions. The South Central region is further broken down into "salt" and "nonsalt" storage facilities, with the majority of the salt caverns existing along the Gulf Coast. Injections (gas added) and withdrawals (gas removed) by region can reflect weather conditions in each area. (Regions were changed in 2015 to better reflect the shift in production coming from the northeastern United States.)

Natural gas storage regions

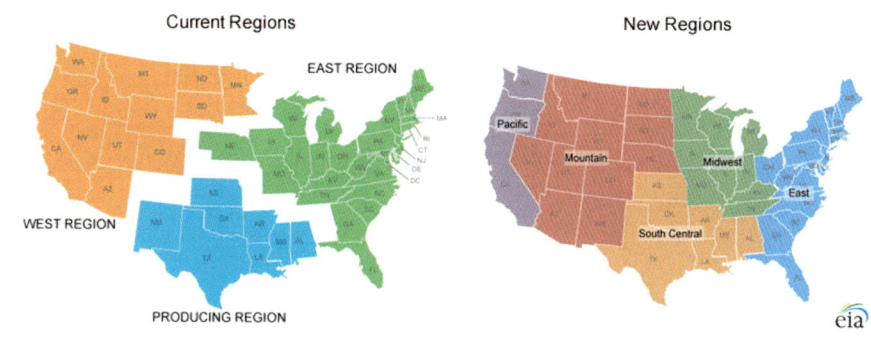

Fig. 1–4. US natural gas storage regions

2. *Total gas in storage.* This is the first thing that traders and others involved in natural gas markets will look to for guidance. It is the change in storage levels from one week to the next. In general, additions to storage (*injections*) are considered a *bearish* price signal; that is, production exceeded demand for the prior week. The converse is true for the removal of gas from storage (*withdrawals*). The indication is that demand exceeded production for the prior week. Prior to the release of the report, analysts have compiled forecasts, and the variance of the actual volume to these predictions causes immediate, sometimes extreme, reactions by traders.

3. *Comparison to "year-ago" data.* This comparison, both in volumetric and percent of variance formats, represents the current inventory level compared with the same period from the prior year. In order to truly interpret this accurately, one must know the prior year's weather, as well as the current weather forecasts.

4. *Comparison to the "5-year average" data.* This comparison is presented in volumetric, percent of variance, and graphic formats. The five-year chart easily demonstrates the state of current storage volumes vs. the past five years for the same time period. The five-year chart, as shown in figure 1–5, clearly indicates the impact of the harsh winter of 2013/2014 on stored volumes throughout that time frame (bottom line). In addition, it was obvious that the volumes in storage as the winter of 2014/2015 began were far less than what would normally be the case. This chart, then, could be viewed as bullish pricing information, especially in light of the prior winter. In actuality, however, by fall 2015, stored volumes surpassed the five-year average, in part because of the natural gas supply surplus. They neared record-high summer ending levels in 2016, only to return to five-year average levels by early 2017.

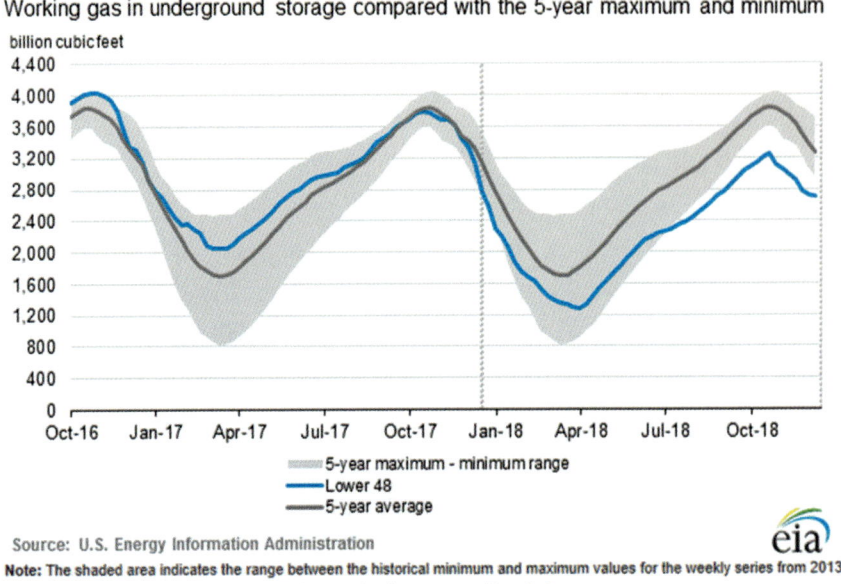

Fig. 1–5. Working natural gas in underground storage compared with the five-year maximum and minimum

Fuel switching

The generation of electricity can be accomplished by using fossil fuels such as natural gas, coal, and fuel oil. In addition, nuclear power plants, wind, solar, and hydroelectric dams can all produce power. The mix of electricity sources thus has a direct impact on natural gas demand and resulting prices. When nuclear power plants are offline due to unexpected outages or for seasonal maintenance and refueling,[30] natural gas–fired power plants tend to take up the slack. When snowpack in the northwestern United States is low during winter, there will be less water run-off from the mountains to the rivers in the spring, leading to lower amounts of potential hydroelectric power available in the summer months. (The exact opposite was true in 2017, when record snowfall and rain amounts resulted in historically high run-off levels.) When the wind does not blow or the sun does not shine, wind generators and solar panels do not produce power. Here again, natural gas–fired plants can easily respond to these swings in electric demand.

The ups and downs of natural gas and coal prices also provide opportunities for electric utilities and independent power producers (IPPs) to save on fuel costs. The low natural gas prices in 2012 and again during the 2014 to 2016 period caused shifts from higher-polluting coal to cleaner-burning, lower-cost natural gas. Stricter emissions standards placed on coal plants during the early 21st century led to more use of natural gas for the plants that could switch fuels. Other plants were retired, leaving the power gap to be filled by natural gas generation. Figure 1–6 illustrates US energy consumption by source. Note the minor amount of energy produced by all sources of alternative and renewable energy sources. Natural gas, with fewer emissions than coal and fewer safety concerns than nuclear power, will have to be the "bridge" fuel until those sources can grow significantly, which could take decades. In fact, in 2016, natural gas overtook coal in terms of the share of power generation.

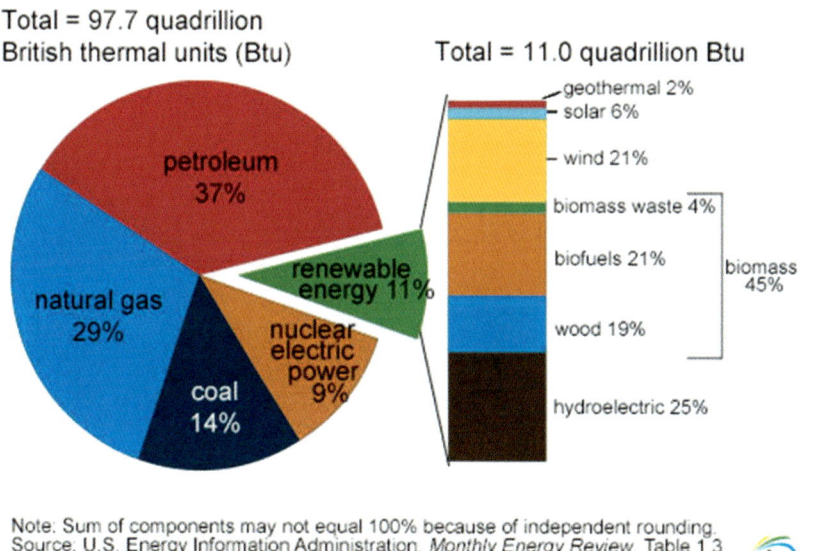

Fig. 1–6. Primary US energy consumption by source and sector, 2017

In this chapter, we explored the many factors that can, and often do, influence the direction of energy commodity prices. In the next chapter, we will look at the physical cash marketplace, where prices are established for energy commodities through transactions representing daily, weekly, monthly, yearly, or even hourly time periods. They call for the actual physical exchange of the commodity.

Summary Points

1. Energy prices can be influenced by a host of factors and the interpretation of them by market participants.
2. Weather in the US is a major factor in price direction for natural gas but a lesser factor for crude oil.
3. Supply & Demand statistics address the actual physical energy commodity balance.
4. Economic indicators infer a future increase/decrease in energy consumption.
5. Geopolitical events involving oil-producing countries or regions can cause concerns over disruption of supply.
6. Periodic reports from both US domestic sources and international groups can provide a wealth of information on the picture of energy production and demand.

Review Exercises

1. Name five factors that can influence the price of crude oil.
2. Name five factors that can influence the price of natural gas.
3. What US government agency provides a vast amount of statistical data on all forms of energy?
4. What is the no. 1 source of energy production in the United States?
5. Which fuel is the most used for production of electricity in the United States?
6. If the US stock market closed higher today, what impact could that have on future energy prices?
   a. Bullish
   b. Bearish
   c. Neutral
7. OPEC has decided to end its production quota curtailments. What impact could that have on future crude oil prices?
   a. Bullish
   b. Bearish
   c. Neutral
8. China's economic growth has slowed. What impact could that have on future crude oil prices?
   a. Bullish
   b. Bearish
   c. Neutral
9. The USD traded lower this week against the *Wall Street Journal* basket of currency indexes. What impact could that have on future crude oil prices?
   a. Bullish
   b. Bearish
   c. Neutral
10. The International Energy Agency (IEA) in Paris has forecasted an increase in demand for oil for next year. What impact could that have on future crude oil prices?
    a. Bullish
    b. Bearish
    c. Neutral
11. The EIA's "Weekly Petroleum Status Report" showed a decrease in refinery utilization last week. What impact could that have on future crude oil prices?
    a. Bullish
    b. Bearish
    c. Neutral

12. The EIA's "Weekly Natural Gas Storage Report" showed a larger-than-forecasted injection for last week. What impact could that have on future natural gas prices?
    a. Bullish
    b. Bearish
    c. Neutral
13. Air-conditioning loads for most of the country are heavy at the same time that six of the nation's nuclear power plants are offline. What impact could that have on future natural gas prices?
    a. Bullish
    b. Bearish
    c. Neutral
14. Snowpack in the northwestern United States was below normal this past winter, and above-normal temperatures have been forecasted for the US West Coast. What impact could that have on future natural gas prices?
    a. Bullish
    b. Bearish
    c. Neutral
15. A new US LNG export terminal has started commercial operations. What impact could that have on future natural gas prices?
    a. Bullish
    b. Bearish
    c. Neutral

Notes

1. We will have to use the term *humans* loosely here, given the advent of supercomputer trade execution or *high-frequency trading*. Despite the electronic execution of trades, there are humans behind the business, with profit as their goal.
2. In reference to crude oil, the term *sweet* means that it has a relatively low sulfur content and is thus less *sour* than other grades. It has been said that the way this was determined in the past was to actually taste the crude.
3. US Energy Information Administration, *Short-Term Energy Outlook Model Documentation: Regional Residential Heating Oil Price Model* (Washington, DC: EIA, 2009), 7, http://www.eia.gov/forecasts/steo/documentation/heating_oil.pdf.
4. National Oceanic and Atmospheric Administration, National Weather Service, "Climate Prediction Center: Degree Days Statistics," http://www.cpc.ncep.noaa.gov/products/analysis_monitoring/cdus/degree_days/.
5. US Energy Information Administration, "Gulf of Mexico Fact Sheet: Overview," http://www.eia.gov/special/gulf_of_mexico/index.cfm.
6. "Special Report: Hurricane Katrina Damage Assessment," *Rig Zone* (September 2, 2005), http://www.rigzone.com/news/article.asp?a_id=24992.
7. National Oceanic and Atmospheric Administration, National Hurricane Center, Home page, http://www.nhc.noaa.gov/.
8. An index above 50 indicates expansion in activity, while lower than 50 indicates contraction; 50 is considered neutral. "Purchasing Managers' Index—PMI," *Investopedia*, http://www.investopedia.com/terms/p/pmi.asp.
9. World Bank Group data as provided by www.tradingeconomics.com.
10. US Energy Information Administration, "China: Overview" (May 14, 2015), https://www.eia.gov/beta/international/analysis.php?iso=CHN.
11. US Energy Information Administration, "China Is Now the World's Largest Net Importer of Petroleum and Other Liquid Fuels," *Today in Energy* (March 24, 2014), http://www.eia.gov/todayinenergy/detail.cfm?id=15531.
12. US Energy Information Administration, "Japan Is the Second-Largest Net Importer of Fossil Fuels in the World," *Today in Energy* (November 7, 2013), http://www.eia.gov/todayinenergy/detail.cfm?id=13711.
13. US Energy Information Administration, "India: Analysis—Energy Sector Highlights," https://www.eia.gov/beta/international/country.php?iso=IND.
14. *BP Statistical Review of World Energy 2014*, 63rd ed. (London: BP, 2014), https://www.bp.com/content/dam/bp-country/de_de/PDFs/brochures/BP-statistical-review-of-world-energy-2014-full-report.pdf.
15. US Energy Information Administration, "United Kingdom: Overview" (as updated March 19, 2018), https://www.eia.gov/beta/international/country.php?iso=GBR.
16. US Energy Information Administration, "Russia: Overview" (as updated October 31, 2017), http://www.eia.gov/beta/international/analysis.cfm?iso=RUS&src=home-b5.
17. US Energy Information Administration, "International Energy Statistics: Total Petroleum and Other Liquids Production," http://www.eia.gov/cfapps/ipdbproject/iedindex3.cfm?tid=5&pid=53&aid=1&cid=RS,SA,&syid=2009&eyid=2013&unit=TBPD.

18. See Office of the Oklahoma Secretary of Energy and Environment, "Earthquakes in Oklahoma: Oklahoma Corporation Commission," https://earthquakes.ok.gov/what-we-are-doing/oklahoma-corporation-commission/.
19. *Bitumen* is the resultant thick crude when tar sands oil is separated from the sand.
20. https://www.usatoday.com/story/news/nation/2015/12/16/oil-exports-ban-congress/77420824/
21. The increase in US oil exports, coupled with vast production increases in Texas, prompted the New York Mercantile Exchange (NYMEX) to initiate a brand new futures contract for the delivery of WTI crude oil to the Houston refinery corridor. This new delivery point at an Enterprise Products facility is deemed to be more reflective of the price impact of the oil export market than the standard, NYMEX WTI-Cushing, OK Hub, futures contract.
22. *BP Statistical Review of World Energy 2014.*
23. Baker Hughes releases updates on the status of active oil and gas drilling rigs. The weekly change is seen by some in the market to indicate supply/demand shift in the future. See Baker Hughes, "Rig Count Overview and Summary Count," http://phx.corporate-ir.net/phoenix.zhtml?c=79687&p=irol-rigcountsoverview.
24. US Energy Information Administration, "Weekly Petroleum Status Report," http://www.eia.gov/petroleum/supply/weekly/.
25. This excludes the *Strategic Petroleum Reserve*, which is the world's largest emergency supply of oil and is controlled by the US Department of Energy. The President can release this supply if needed via competitive sales. Even the prospect of a sale can directly influence the price of oil in North America. Total capacity now sits at 727 million bbl. See US Department of Energy, Office of Fossil Energy, "Strategic Petroleum Reserve," http://energy.gov/fe/services/petroleum-reserves/strategic-petroleum-reserve#Current.
26. The first shipment of LNG from Cheniere Energy's Sabine Pass LNG export terminal occurred in February 2016. As volumes from this terminal increase and others near completion, North American natural gas will enter the global LNG market in a big way. See "Sabine Pass LNG Terminal," https://www.cheniere.com/terminals/sabine-pass/.
27. The meteorological term for this storm phenomenon is actually *bombogenesis*.
28. As of this writing, the only two export facilities operating were Cheniere's Sabine Pass, LA, and Dominion Energy's Cove Point, MD, totaling about 3.5 bcfd of LNG exports.
29. The report can be found at US Energy Information Administration, "Weekly Natural Gas Storage Report," http://ir.eia.gov/ngs/ngs.html.
30. The US Nuclear Regulatory Commission has a daily status report for all US nuclear power plants. See US Nuclear Regulatory Commission, "Power Reactor Status Report," http://www.nrc.gov/reading-rm/doc-collections/event-status/reactor-status/ps.html.

2

Oil and Natural Gas Cash Markets

Key Learning Points

- Cash price postings ("indexes") reflect physical commodity trading.
- The physical commodity market has its own pricing scheme.
- Cash market prices are reported by industry publications using survey methodology and are known as *indexes* or *postings*.
- There are some key industry publications for oil and gas pricing, and it will be helpful to be familiar with these resources and their uses:
- *Platts Gas Market Report*
- *Platts Gas Daily*
- *OPIS* price reports
- *ARGUS* crude reports
- Online price data resources such as the Intercontinental Exchange (ICE) are useful tools for observing physical, cash market prices.
- *Platts Gas Market Report* and *Platts Gas Daily* are the main postings for natural gas.
- OPIS (Oil Producers Information Service) reports are the primary reports showing postings for US crude and natural gas liquids (NGLs).
- *ARGUS* reports global oil prices and is a widely used posting for crude oil contracting.
- ICE is a totally electronic, over-the-counter platform trading thousands of physical and financial products. Daily "settlement" prices are published for cash transactions.[1]

Energy is being consumed every hour of the day, everywhere on earth. Thus, energy commodities are being bought and sold constantly to fill this demand. When we are talking about prices for the actual *physical* production and consumption of natural gas and crude oil, we are talking about the cash or spot market, where "transactions [are] for immediate delivery."[2] While the primary focus of this book is on the financial and not, the physical energy markets, the two are inextricably intertwined.

We must also differentiate between the derivatives *futures* and *forwards*, as the latter involves the physical market, while the former involves the financial market. Forward contracts require delivery of the commodity and are traded in the OTC markets.

Forwards can be defined as "a bilateral transaction that provides for delivery of the underlying commodity at a future date, at the price determined at the inception of the contract, with cash being paid at delivery, or after an agreed number of days following the delivery."[3] Or, more simply put, "It is an agreement to buy or sell an asset at a certain future time for a certain price."[4] Mack (2014) would add that the price is agreed to in the present time, stating that "a forward contract is a nonstandardized contract between two parties to buy or sell an asset at a specified future time at a price agreed upon today."[5] They are bilateral agreements between the two parties that are not exchange traded. Mack further postulates that forward contracts can be physical or financial, depending on when they are settled, stating:

- If a forward contract is settled before its maturity date, it is a *financial forward contract* since no... [commodity] is physically delivered.
- A forward contract is a *physical contract* if the...[commodity] is delivered physically.[6]

Forward contracts are different than *spot contracts*, which are agreements to buy or sell the commodity almost immediately. According to Mack, "The spot market is a commodities or securities market in which goods are sold for cash and delivered immediately. Contracts bought and sold on these markets are immediately effective."[7]

However, the exact differences between the two often become lost in usage. Any agreement to exchange a commodity at a future point in time can be considered a "forward" arrangement, so Kaminski makes the following points:

- The demarcation line between spot and forward transactions is, at best, blurred;
- from the practical point of view, the forward markets are critical for energy trading—and this is what counts;
- the price formation process in what one can call the spot markets is increasingly dependent the price discovery in the forward and futures transactions;
- practical difficulties of identifying true spot prices leads most researchers to using the futures prices of the shortest maturity as the proxy for spot; and,
- the elusive nature of spot prices does not stop many modellers from building models using spot prices as the cornerstone of their philosophical approach.[8]

Fundamental supply/demand changes in the physical marketplace impact the cash market prices but also provide signals to the financial markets. Conversely, changes in the prices in the financial markets influence cash markets and provide signals as to future prices. But we need to make sure that we understand the distinction between the two where the time frame is concerned. Financial markets deal with prices beginning at some future point in time, normally the next "whole" month. Cash market prices deal with current supply and demand conditions. They represent the value of natural gas, oil, and natural gas liquids in the here and now. (In a later chapter, we will deal with the idea of "fixed" prices in the cash market for longer-term transactions. But setting these prices still depends on the future markets. When we speak of cash markets being "in the present," we are dealing with the day-to-day changes in price that cannot be set using the financial market derivatives.)

In this chapter, we will explore the ways in which cash prices are established in the physical marketplace for oil, natural gas, and natural gas liquids (NGLs), the main media sources that report these prices, and the methodologies they use to collect the data.

Natural Gas and Crude Oil—Physical Pricing

Even though the prices of energy futures influence the physical markets, cash prices are negotiated outside the confines of the formal exchanges. Buyers and sellers, looking at their supply and demand situations, make pricing decisions daily, and buy and sell the *physical* commodities. The results of these trades are reported in industry publications and become market indicators for the physical cash market.

As one can probably guess, variations in supply and demand for energy are constantly occurring. Dramatic changes in demand for oil do not occur as frequently as with natural gas, which is largely weather-driven. This means that the exchange of energy from one party to another is ongoing, that is, negotiations are taking place throughout each day for various volumes and time frames. Furthermore, as natural gas continues to increase in its share of power generation as coal declines, demand for natural gas can change hourly.

Negotiations for increased or decreased volumes of natural gas and crude oil can take place monthly, weekly, daily, and even "intraday," in order to balance the market's supply and demand status. These transactions, or *deals* as they are known, have to be priced somehow. In addition, the prices generally take two forms: fixed or indexed.

Fixed prices are set prices agreed upon by the negotiating parties for a given volume of energy changing hands. Factors such as futures prices and other prices being traded at the time are taken into consideration when establishing these prices.

These days, the vast majority of fixed-price transactions for physical oil and natural gas are conducted on electronic trading platforms such as the ICE, one of the world's largest online trading platforms.[9] Few physical energy commodity trades are consummated over the phone these days, as counterparties tend to rely more on electronic trading systems and various forms of instant messaging.

The results of these cash market transactions are reported to industry publications on a voluntary basis. Reporters request the volumes and prices from the various counterparties to these deals. From there, volume and pricing information is calculated relative to the specific physical location where delivery of the commodity is to take place. Most publications use what is known as a *weighted average price* to determine what price to publish for each location. In other words, instead of taking a simple average of prices reported, the volume attached to each price is weighted relative to other volumes and prices reported.

Table 2–1 gives an example of how these calculations work. Note the difference between the simple average of the reported prices and the weighted average. The latter includes consideration of the volume associated with each price. ("Weighted Average Sales Price (WASP)" for those selling natural gas; Weighted Average Cost of Gas (WACOG)" for those buying natural gas.)

Table 2–1. Example of weighted average price calculation

	Price	Volume MMBtu	
Transaction 1	$3.240	2,500	$8,100
Transaction 2	$3.210	7,500	$24,075
Transaction 3	$3.220	10,000	$32,200
Transaction 4	$3.280	4,500	$14,760
Transaction 5	$3.250	5,000	$16,250
Transaction 6	$3.200	10,000	$32,000
Transaction 7	$3.280	12,000	$39,360
Transaction 8	$3.290	8,000	$26,320
Transaction 9	$3.300	3,000	$9,900
		62,500	$202,965
Simple Avg.	$3.252		
Weighted Avg.			$3.247

The publications then post all prices in their respective periodic newsletters. The prices are then recognized as the postings or indexes for the location and time frame shown. Indexes comprise the second generally used form of deal pricing. Rather than haggle over what fixed price to use, counterparties can merely agree to settle their transactions using the price or prices that are published in the relevant newsletter. Purchases and sales using these indexes for settlement are deemed to be at "market" prices. For those responsible for buying natural gas, crude, or NGLs, conducting business using these accepted postings is considered to be a prudent purchasing method. Sellers of these same commodities accomplish their objectives by selling at the market level.

Key publications for the respective commodities indicated include the following:

- McGraw-Hill: *Platts Gas Market Report*. This newsletter posts the fixed prices for physical natural gas transactions that cover an entire calendar month.[10]
- McGraw-Hill: *Platts Gas Daily*. This newsletter posts fixed prices for physical natural gas transactions that are for a duration of one day (or weekend) only.[11]
- OPIS (Oil Producers Information Service). OPIS publishes the primary report showing postings for daily North American natural gas liquids (NGL) prices.[12]
- ARGUS. Argus Media reports provide hundreds of price locations for North American and international crude oil and refined products.[13]

In addition, ICE posts the daily settlement prices for several energy commodity prices, including oil and natural gas delivery points throughout North America, based upon actual trading on their platform.[14]

The Intercontinental Exchange (ICE) Trading Platform

In figure 2–1 there is an example of an ICE trading screen for physical natural gas. The column labels are defined in the following text.

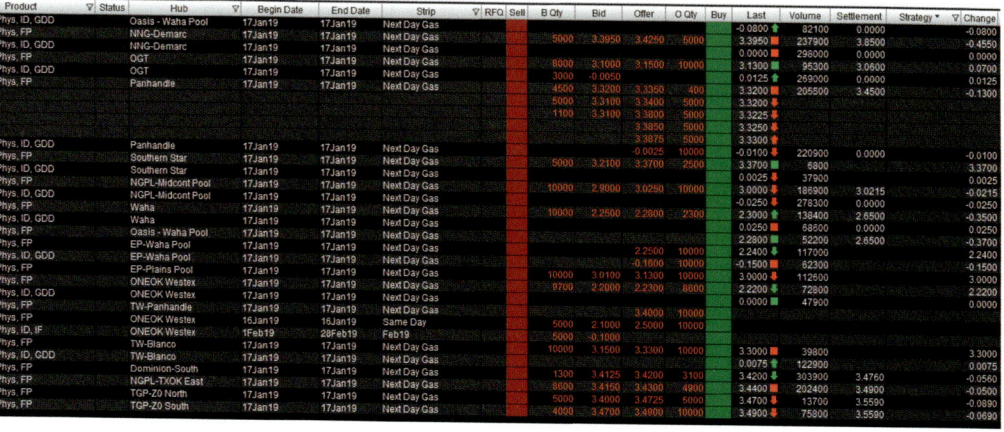

2–1. Daily physical natural gas trading screen from the Intercontinental Exchange

Column labels

Product. The instrument or commodity trading. In this case, they are all "NG Firm Phys" for physical natural gas bought/sold on a *firm* basis, which means both parties are obligated to perform once the trade is executed, barring any unforeseen event of *force majeure*. The pricing to be used is specified next. "FP" simply denotes a fixed price. (These are transactions that may get reported to the natural gas price publications.) "ID" is an abbreviation for index and indicates that the volumes traded will be priced at the referenced publication's posting. Following the index abbreviation is the designation for the publication. "GDD" stands for the daily index from *Gas Daily*, which is now known as *Platts Gas Daily*. At the bottom of the screenshot, you will see a different index indicator. "IF" stands for *Inside FERC*, the old name for what is now *Platts Gas Market Report*.

Hub. This identifies the pipeline delivery point, which is also associated with the respective index/posting.

Strip. The time frame for the exchange of the natural gas. In this case, "Next Day Gas" indicates that the transaction will be for the next *gas day*, a period running from 9:00 AM central time zone one day to 9:00 AM central time zone the next day, the standard 24-hour pipeline flow measurement cycle. In this case, buyers and sellers of natural gas are transacting business based upon supply available and demand forecasted, for the next day only. Again, notice the transaction information at the bottom of the screenshot. The strip in this case is for the entire period of November 2016 through March 2017, "Nov16–Mar17."

Begin date. Initial date of production/gas flow.

End date. Ending date of production/gas flow. Note that for "Next Day Gas," the begin date and end date are the same, whereas for "Nov16–Mar17," the begin date is "1Nov16" and the end date is "31Mar17," encompassing what the industry regards as the winter months.[15] This, then, in the vernacular of energy trading, is the *winter strip*.

+. The plus symbol indicates that there are more possible trades or "depth" available for view. By clicking on the symbol, all other bids and offers are shown. When collapsed, only the best bid and offer will appear. ("NG Firm, Phys Panhandle" has been expanded for illustrative purposes.)

B Qty. The quantity (volume) of natural gas that corresponds to the bid price.

Bid. The price the buyer is willing to pay for the volume indicated.

Offer. The price the seller is willing to take for the volume indicated.

O Qty. The volume of natural gas that corresponds to the offer price.

Last. The price of the last trade that was executed.

Change. The difference in price between the "Last Trade" and the prior day's "Settlement" price.

Settlement. The prior day's weighted average price for transactions at the location. For FP transactions, this is a fixed-price, while for ID trades, it can represent a premium paid or deduction made from the published index.

High. The highest price traded thus far on the day.

Low. The lowest price traded thus far on the day.

WAP. During active trading, this represents a running weighted average price for all trades at the location. When trading finishes, this becomes the settlement for this day.

Volume. The total volume of natural gas traded at the location and price indicator.

Summary Points

1. Cash prices reflect physical commodity trading.
2. Posted prices or indexes represent the market price.
3. Publications poll market participants and calculate weighted average prices.
4. Indexes are posted daily, weekly, and monthly.
5. Two main price publications exist for natural gas.
6. Two main price publications exist for crude oil and natural gas liquids.
7. ICE publishes daily settlement prices from its electronic trading platform for oil and natural gas.
8. The US Energy Information Administration has extensive current and historical market pricing for natural gas, crude oil, and natural gas liquids, available at www.eia.gov.

In this chapter, we addressed the physical cash marketplace that, for the most part, deals with the "here and now." In the next chapter, we will delve into the financial futures markets, whereby fixed (or set) commodity prices can be obtained for future months and years.

Review Exercises

1. What is a forward contract?
2. What does ICE stand for?
3. What publication covers prices for physical natural gas transactions that cover an entire calendar month of production?
4. What publication covers prices for physical natural gas transactions that are for the "next day" of gas flow?
5. What publication is best known for its NGL prices?
6. What publication is best known for its oil prices?
7. What is the name given to the methodology whereby the named publications gather their pricing information?
8. What does WASP mean?
9. What is a WACOG?
10. What is the term given to the last week of each month where trading for the ensuing month occurs?

Notes

1. These can be found at *NGI's Price Alert*, http://www.naturalgasintel.com/ICE.
2. Vincent Kaminski, *Energy Markets* (London: Risk Books, 2013).
3. Ibid.
4. John C. Hull, *Options, Futures, and Other Derivatives*, 9th ed. (Upper Saddle River, NJ: Pearson Education, 2015).
5. Iris M. Mack, *Energy Trading and Risk Management: A Practical Approach to Hedging, Trading and Portfolio Diversification* (Singapore: John Wiley & Sons, 2014), 3.
6. Ibid., 4.
7. Ibid., 7.
8. Kaminski, *Energy Markets*.
9. More information about the Intercontinental Exchange can be found at the ICE website, www.theice.com.
10. *Platts Gas Market Report* is available by subscription at http://www.platts.com/products/gas-market-report.
11. *Platts Gas Daily* is available by subscription at http://www.platts.com/products/gas-daily.
12. OPIS reports for natural gas liquids are available by subscription at http://www.opisnet.com/.
13. Argus media reports for various energy markets are available by subscription at http://www.argusmedia.com/.
14. See *NGI's MidDay Price Alert*, available by subscription at http://www.naturalgasintel.com/ICE.
15. Since energy demand has traditionally been the highest in the winter, the industry chose the months of November through March to represent that period. "Core" winter is considered December/January/February. And thus "summer" is comprised of all the remaining months, regardless of whether or not they fit what we normally consider to be summer months.

3

Financial Energy Commodity Markets

Key Learning Points

- Understand the term *energy derivative*.
- Determine the difference between the physical cash market and financial energy futures.
- Differentiate between exchange-traded and over-the-counter energy derivatives.
- Understand the difference between a forward contract and a futures contract.
- Name the requisite components of a NYMEX futures contract.
- Explain the price discovery provided by a formal futures exchange.
- Compare/contrast futures exchanges with "over-the-counter" (OTC) markets.

Before we begin our discussion of financial energy commodities, it is important to understand what exactly constitutes a commodity. Gérard Debreu, who won the Nobel Prize in economics in 1983, defined *commodity* as follows:

> A commodity is characterized by its physical properties, the date at which it will be available, and the location at which it will be available. The price of a commodity is the amount which has to be paid now for the (future) availability of one unit of that commodity.[1]

According to Kaminski, "Understanding derivative instruments is a critical skill that any participant in the energy markets requires."[2]

The focus of this book is on energy derivatives. Mack defines an energy derivative as "a derivative whose underlying asset is some type of energy product, such as oil, natural gas, or electricity."[3] These can be futures, swaps, spreads, or options.[4] They are used both by speculators hoping to make a profit and by organizations that desire a hedge against fluctuations in pricing. The key here is that they *derive their value* from an actual underlying physical commodity. And it is the price of that underlying commodity that transcends all layers of derivative products.

Financial markets for energy commodities are comprised of two main groups: exchange-traded and over-the-counter (OTC). Exchange-traded markets are "a centralized market where buyers and sellers, through brokers, interact and trade financial energy products."[5] Hull states, "A derivatives exchange is a market where individuals trade standardized contracts that have been defined by the exchange."[6] These markets provide credit assurances that guarantee contract execution, performance, and settlement. Transactions can occur in the "pits" on the trading floor of an exchange (*open outcry*), by phone brokers, and electronically. The volume of trading conducted on the floors of the exchanges, however, is declining.

Some of the advantages that formal exchanges offer are the following:

- *Standardization.* Uniform volume and quality requirements reduce variations in contract terms. Variations can add costs due to the number of transactions that may be required to make necessary adjustments. Standardization allows clients to simply pay a fee to their brokers, who pay the exchange a clearing fee that guarantees the transaction.
- *Reduced credit risk.* While clients post some form of credit assurance to their broker, the broker has the obligation to provide the same to the exchange as well. In some cases, the exchanges act as the clearing party and require cash deposits, known as *margins*, from trading parties.
- *Price discovery.* The market prices of completed transactions are conveyed within minutes across the globe for everyone to see. This visibility provides all participants the same information at the same time, thus theoretically offering a level playing field.
- *Liquidity.* Transactions are negotiated with hundreds of counterparties, making it easier to obtain the price, volume, and term needed.

Non-exchange OTC markets can take the form of *voice brokers*, where orders are literally conducted by phone or, through electronic platforms such as the ICE and CME Group's *ClearPort*. Main players there include banks, fund and equity managers, large financial institutions, and corporations. OTC markets, while providing customized contracts, lack some of the advantages of the formal exchanges. Price transparency, in the case of phone transactions, is practically nonexistent unless one has access to real-time market data. Credit assurances have to be coordinated between the two actual counterparties to the agreement, either through the use of a central counterparty (CCP) or through the terms and conditions of their bilateral agreements. Transaction fees can mount up as multiple deals may have to be completed with multiple counterparties. Liquidity can be lacking as OTC markets may not have as many participants. (OTC markets such as ICE and *ClearPort*, however, now have thousands of counterparties.) Institutions that are willing to provide both a bid and an offer for derivatives transactions are known as *market makers* and take on the risk of the trade. Nevertheless, their bid/ask spread usually reflects some margin for them, or they can lay off their risk with another counterparty or even by trading other financial instruments in their overall "book."

Forward contracts are traded in OTC markets as an alternative to futures. According to Hull, they represent "an agreement to buy or sell an asset at a certain future time for a certain price," with physical settlement.[7] Kaminski defines a forward contract as "a bilateral transaction that provides for the delivery of the underlying commodity at a future date, at the price determined at the inception of the contract, with cash being paid at delivery, or at an agreed upon number of days after delivery."[8] And Mack defines a forward contract as "a nonstandardized contract between two parties to buy or sell an asset at a specified future time at a price agreed upon today."[9] Furthermore, if a forward contract is settled before the maturity date of the contract, it is considered to be a *financial forward contract*. Conversely, if the commodity is delivered, the forward contract is a *physical forward*. Forward contracts are bilateral agreements between a buyer and a seller and are not exchange traded.

One party to the forward contract takes a *long position*, agreeing to buy the asset at the agreed-upon price and date, while the other party takes a *short position*, agreeing to sell the asset at the same price and date. Essentially, any purchase or sale of an asset at a future point in time would be a forward transaction. The physical spot market for energy commodities would then fall under this category, with the possible exception of real-time power trading, which calls for immediate delivery.

Comparison of Futures vs. Forward Contracts

Futures	Forwards
Exchange traded	Private negotiation
Standardized	Nonstandard, customized
Multiple delivery dates	Single delivery date
Daily settlement(s)	End of contract settlement
Position normally closed out before/on expiration	Delivery or cash settled
Exchange-backed credit assurances	Counterparty credit risk

Energy Futures Contracts

...an agreement to buy or sell an asset at a future time for a certain price.

—John C. Hull

Since the crux of this book deals with financial energy derivatives, we must start with the primary underlying contract—futures. As mentioned above, these contracts are not dealing with spot or cash markets. They represent the future purchase or sale of a specific energy commodity, at an agreed-upon price, for a specified time frame and delivery location. The spot market entails energy commodity transactions calling for cash payments and "immediate"[10] delivery. According to the Global Association of Risk Professionals,

> "Futures are forward contracts traded on an organised exchange that offer credit guarantees of counterparty performance, provide trading infrastructure (typically an open outcry system or screen- based electronic trading) and develop the definitions of standardised contracts."[11]

Mack states,

> "A futures contract is a standardized contract between two parties to buy or sell a specified asset of standardized quantity and quality for a price agreed upon today with delivery and payment occurring at a specified future date. The contracts are negotiated at a futures exchange, which acts as an intermediary between the two parties."[12]

Inkpen and Moffett offer the following description of the futures market from NYMEX:

> "The futures markets help businesses manage their price risk by providing a means of hedging; matching buyers and sellers of a commodity with parties who are either more able to bear market risk, or who have inverse risk profiles....
>
> Because futures are traded on exchanges that are anonymous to public auctions with prices displayed for all to see, the markets perform the important function of price discovery....They reflect the marketplace's collective valuation of how much buyers are willing to pay and how much sellers are willing to accept. The diverse views of many market participants are distilled to a single price."[13]

ICE Futures Europe[14] is a major European electronic trading platform, second only to NYMEX in terms of commodity futures trading. ICE Futures Europe further describes futures:

> "Futures prices are not price predictions but are the collective current opinion of the marketplace of where prices appear to be heading. That opinion, and the direction of prices, can change in an instant, which makes these markets so challenging and potentially rewarding."[15]

Here again, we see that prices are determined in and by the market, which is comprised of human beings who may react emotionally or even sometimes irrationally.

For our purposes, we will use the definition that is specific to energy futures contracts traded on the New York Mercantile Exchange (NYMEX): "A legally binding obligation for the holder of the contract to buy or sell a particular commodity at a specific price and location at a specific date in the **future**."[16]

Within this definition, the basic elements of all commodity futures contracts are presented:

1. *Commodity*. The type of energy to be bought or sold.
2. *Price*. The agreed-upon market price for the commodity (perhaps the most important element).
3. *Location*. The place where the commodity is to be exchanged, unique to each contract.
4. *Date*. The *future* month when the exchange will take place. Volumes committed under a futures contract *must* be delivered within the respective month(s) stated.

These contracts are known as *derivatives* because they actually derive their value from some other source. Hull defines a derivative as "a financial instrument whose value depends on (or derives from) the value of other, more basic, underlying variable. Very often the variable underlying derivatives are the prices of traded assets."[17] In our case, the futures contracts derive their value from the underlying energy commodity applicable to each one. These also represent the first building block in the more advanced financial energy derivatives we will discuss in a later chapter.

We need to emphasize the legal nature of these contracts. They are *binding obligations* to "make or take" the commodity. Thus these are firm transactions, and failure to perform under the agreements may result in legal and financial ramifications, both with the counterparty to the transaction and with the exchange itself. Additionally, consequences such as civil penalties and fines imposed by jurisdictional regulatory bodies could occur, depending on the circumstances of the default.

It should be stressed that each of these futures contracts represents the actual underlying commodity. They represent the obligation to buy or sell *actual* oil, natural gas, gasoline, or heating oil. The vast majority of contracts are traded strictly for financial purposes (according to the CME, fewer than 2% of all contracts traded result in actual physical delivery). The idea that each contract represents the actual commodity tends to get lost when one views the market's activity. This is never truer than when those who do not understand the financial energy markets describe the prices as being the result of "smoke and mirrors" caused by speculators. Quite the contrary, when futures contracts expire, those holding them are obligated to deliver or receive the actual commodity at the designated delivery point. Often noncommercial traders (speculators) find themselves with an unbalanced book at the end of a month, which requires them to become involved in the physical marketplace to square their positions. This was certainly not their intention when they entered into those transactions.

Futures perform several important functions. Prices, which are determined in a highly competitive market, are reflective of that market's *perception*, at the time, of the future value of the commodity based on supply and demand dynamics. *Price discovery* is provided as those prices are relayed almost instantaneously to financial information sources across the globe, such as news sources, websites, electronic platforms, etc. This visibility allows all interested parties the same advantage in negotiating futures transactions. Futures also provide an actual marketplace for commercial entities that buy and sell the physical commodities, allowing them opportunity to reduce their price and market or supply risk, otherwise known as *hedging*. (We will cover hedging in a later chapter.) Counterparty credit risk is reduced, since the exchange guarantees the trades. Furthermore, trading is conducted by hundreds of parties, making liquidity[18] higher.

Summary Points

1. There are (2) distinct but interrelated markets for energy commodities, physical and, financial.
2. Financial derivative contracts "derive" their value from the underlying energy asset (commodity).
3. NYMEX is a formal, regulated exchange with standardized contracts for financial energy commodities.
4. NYMEX energy commodity futures provide much-need "price discovery" for all market participants and interested parties.
5. "Over-the-counter" (OTC) markets also exist for physical and, financial, energy commodities.
6. Futures are traded on a regulated exchange while forwards are traded in the OTC market.

Review Exercises

1. What is a forward contract?
2. What is a futures contract?
3. Name the required elements of a NYMEX futures contract.
4. List four advantages of futures contracts vs. forward contracts.
5. What does OTC mean?
6. What is NYMEX's OTC exchange called?
7. What OTC exchange exists for trading in Europe?
8. What is a derivative contract?
9. What is the underlying asset for an energy derivative contract?
10. What does NYMEX stand for?

Notes

1. McDonald, R. L., *Derivatives Markets*, 3rd. ed.(London: Prentice Hall, 2013)
2. Vincent Kaminski, *Energy Markets* (London: Risk Books, 2013).
3. Iris M. Mack, *Energy Trading and Risk Management: A Practical Approach to Hedging, Trading and Portfolio Diversification* (Singapore: John Wiley & Sons, 2014).
4. Ibid.
5. GARP, *.Foundations of Energy Risk Management.* (Hoboken, NJ: Wiley)
6. John C. Hull, *Options, Futures, and Other Derivatives*, 9th ed. (New York: Pearson Education, 2015).
7. Ibid.
8. Kaminski, *Energy Markets*.
9. Mack, *Energy Trading and Risk Management*.
10. Spot market sales and purchases can be for "same-day," "next-day," next week, or even next month. But they differ from futures contracts in that they involve a more "immediate" delivery.
11. Global Association of Risk Professionals.
12. Mack, *Energy Trading and Risk Management*.
13. Andrew Inkpen and Michael Moffett, *The Global Oil and Gas Industry*: Management, Strategy, and Finance (Tulsa: PennWell, 2011), 381.
14. A part of the Intercontinental Exchange (ICE).
15. Inkpen and Moffett (2011).
16. CME Group (2014).
17. Hull (2015)
18. The larger the number of potential trading counterparties, the greater the chance of being able to get the quantity desired at a competitive price. Liquidity risk can be an issue in OTC markets.

4

The New York Mercantile Exchange

Key Learning Points

- Understand the history and development of the New York Mercantile Exchange.
- Know the energy commodities traded.
- Know the difference between pit and electronic trading.
- Understand the specific futures contract provisions for the following:
  - Natural gas
  - Crude oil
  - Heating oil
  - Unleaded gasoline
- Understand the importance of the price discovery function provided by NYMEX for energy commodities.
- Know the two NYMEX electronic trading platforms.
- Become familiar with the operating hours of the NYMEX.
- Identify NYMEX market participants by type.

Futures exchanges provide the means for those who wish to buy or sell assets in the future to consummate those trades in the present time. According to Errera,

"The prices of commodity futures contracts are determined in a highly efficient central marketplace and at any point in time prices reflect the market's best estimate of the correct price of the commodity given all of the factors that are known to have an impact on the current and future level of energy prices."[1]

Futures markets have existed as far back as the Middle Ages.[2] The New York Mercantile Exchange has been around since the late 1800s in the United States, and it is still the most influential financial energy commodities exchange in the world. In this chapter, we will explore the history of NYMEX, how it functions, the participants, the commodities traded, and its contract specifications.

History of the New York Mercantile Exchange

Early markets for agricultural goods in large cities were scattered, so the need existed for some form of common marketplace. In addition, farmers and producers had begun to seek a price for their goods in advance of their upcoming harvest or production cycles. As a result, a group of savvy entrepreneurs decided to create a centralized market in the form of an "exchange," where buyers and sellers could meet to conduct transactions for future business.

In 1872, a group of dairy and cheese producers in New York City founded the Butter and Cheese Exchange of New York, whose goal was to

> *"provide some common ground on which all could stand, establish rules and usages that are needed to guard and protect the interest of all, permanently form a unity of general principles and an authority of which the trade is now destitute and which is so greatly required."*[3]

This would later become the Butter, Cheese, and Egg Exchange of New York in 1880. In 1882, the name was changed to the New York Mercantile Exchange or NYMEX, as it is still known today. Products at that point included onions, apples, potatoes, plywood, and platinum. (The last product is the only one still traded on NYMEX.)

It would be 100 years until the world's first energy commodity futures contract emerged, with the addition of heating oil in response to the energy crises of the 1970s.[4] A series of other key energy contracts would be added over the ensuing 12 years. Four energy commodities—heating oil, crude oil, unleaded gasoline, and natural gas—will be the focus of this book.

Contract Inaugural Date	Commodity/Trading Symbol
1978	Heating oil (HO)
1983	Crude oil (CL)
1984	Unleaded gasoline (HU/RBOB)
1990	Natural gas (NG)

In 1994, NYMEX and the Commodities Exchange (COMEX) merged, creating two divisions under the New York Mercantile Exchange. Originally founded as a not-for-profit entity, NYMEX went public in 2006 as NYMEX Holdings, Inc. In 2008, the new entity was acquired by the CME Group and remains the world's largest energy and metals trading platform. Today, the CME Group is comprised of the Chicago Mercantile Exchange (CME), Chicago Board of Trade (CBOT), NYMEX, and COMEX.

NYMEX is a highly regulated marketplace with oversight from both the US Commodities Futures Trading Commission (CFTC) and the NYMEX Board and Compliance staff. In addition to its floor trading, NYMEX also has two electronic platforms, *Globex* and *ClearPort*. Globex is the successor to the ACCESS system, launched in 1993, which was designed to provide futures trading beyond the regular pit sessions. Today, Globex trading occurs from Sunday evening at 6:00 PM (EST) through Friday at 5:15 PM (EST), except for a 45-minute break each trading day from 5:15 PM to 6:00 PM. (It should be noted that trades on Globex executed after the regular session hours are considered effective for the next trading (business) day and are counted in all pricing and volume reported for that session.)

NYMEX Contract Specifications

One of the distinct advantages of regulated exchanges such as the NYMEX is the standardization of the contracts. This uniformity ensures that all transactions are equal in nature. It is imperative for anyone dealing in NYMEX futures contracts to understand these specifications, especially as regards price, delivery location, and the expiration date.

For the commodities we are dealing with, the CME Group has the contract specifications shown in table 4–1.

Table 4–1. CME contract specifications

Crude Oil (CL) – Cushing, OK Hub

Product Symbol	CL	
Venue	CME Globex, CME ClearPort, Open Outcry (New York)	
Hours (All Times are New York Time/ET)	CME Globex	Sunday - Friday 6:00 p.m. - 5:15 p.m. New York time/ET (5:00 p.m. - 4:15 p.m. Chicago Time/CT) with a 45-minute break each day beginning at 5:15 p.m. (4:15 p.m. CT)
	CME ClearPort	Sunday – Friday 6:00 p.m. – 5:15 p.m. (5:00 p.m. – 4:15 p.m. Chicago Time/CT) with a 45-minute break each day beginning at 5:15 p.m. (4:15 p.m. CT)
	Open Outcry	Monday – Friday 9:00 AM to 2:30 PM (8:00 AM to 1:30 PM CT)
Contract Unit	1,000 barrels	
Price Quotation	U.S. Dollars and Cents per barrel	
Minimum Fluctuation	$0.01 per barrel	

Maximum Daily Price Fluctuation	Initial Price Fluctuation Limits for All Contract Months. At the commencement of each trading day, there shall be price fluctuation limits in effect for each contract month of this futures contract of $10.00 per barrel above or below the previous day's settlement price for such contract month. If a market for any of the first three (3) contract months is bid or offered at the upper or lower price fluctuation limit, as applicable, on Globex it will be considered a Triggering Event which will halt trading for a five (5) minute period in all contract months of the CL futures contract, as well as all contract months in all products cited in the Associated Products Appendix of rule 200.06. Trading in any option related to this contract or in an option contract related to any products cited in the Associated Products Appendix which may be available for trading on either Globex or on the Trading Floor shall additionally be subject to a coordinated trading halt.
Termination of Trading	Trading in the current delivery month shall cease on the third business day prior to the twenty-fifth calendar day of the month preceding the delivery month. If the twenty-fifth calendar day of the month is a non-business day, trading shall cease on the third business day prior to the last business day preceding the twenty-fifth calendar day. In the event that the official Exchange holiday schedule changes subsequent to the listing of a Crude Oil future, the originally listed expiration date shall remain in effect. In the event that the originally listed expiration day is declared a holiday, expiration will move to the business day immediately prior.
Listed Contracts	Crude oil futures are listed nine years forward using the following listing schedule: consecutive months are listed for the current year and the next five years; in addition, the June and December contract months are listed beyond the sixth year. Additional months will be added on an annual basis after the December contract expires, so that an additional June and December contract would be added nine years forward, and the consecutive months in the sixth calendar year will be filled in. Additionally, trading can be executed at an average differential to the previous day's settlement prices for periods of two to 30 consecutive months in a single transaction. These calendar strips are executed during open outcry trading hours.
Settlement Type	Physical
Delivery	Delivery shall be made free-on-board ("F.O.B.") at any pipeline or storage facility in Cushing, Oklahoma with pipeline access to Enterprise, Cushing storage or Enbridge, Cushing storage. Delivery shall be made in accordance with all applicable Federal executive orders and all applicable Federal, State and local laws and regulations. At buyer's option, delivery shall be made by any of the following methods: (1) by interfacility transfer ("pumpover") into a designated pipeline or storage facility with access to seller's incoming pipeline or storage facility; (2) by in-line (or in-system) transfer, or book-out of title to the buyer; or (3) if the seller agrees to such transfer and if the facility used by the seller allows for such transfer, without physical movement of product, by in-tank transfer of title to the buyer.

Delivery Period	(A) Delivery shall take place no earlier than the first calendar day of the delivery month and no later than the last calendar day of the delivery month. (B) It is the short's obligation to ensure that its crude oil receipts, including each specific foreign crude oil stream, if applicable, are available to begin flowing ratably in Cushing, Oklahoma by the first day of the delivery month, in accord with generally accepted pipeline scheduling practices. (C) Transfer of title-The seller shall give the buyer pipeline ticket, any other quantitative certificates and all appropriate documents upon receipt of payment. The seller shall provide preliminary confirmation of title transfer at the time of delivery by telex or other appropriate form of documentation.

Source: CME Group, "Crude Oil Futures Contract Specs" (2019).

Natural Gas (NG) – Henry, LA Hub

Code	NG	
Venue	CME ClearPort, CME Globex, Open Outcry (New York)	
Hours (All Times are New York Time/ET)	CME Globex:	Sunday – Friday 6:00 p.m. – 5:15 p.m. (5:00 p.m. – 4:15 p.m. Chicago Time/CT) with a 45-minute break each day beginning at 5:15 p.m. (4:15 p.m. CT)
	CME ClearPort:	Sunday - Friday 6:00 p.m. - 5:15 p.m. (5:00 p.m. - 4:15 p.m. Chicago Time/CT) with a 45-minute break each day beginning at 5:15 p.m. (4:15 p.m. CT)
	Open Outcry:	Monday – Friday 9:00 a.m. – 2:30 p.m. (8:00 a.m. – 1:30 p.m. CT)
Contract Unit	10,000 million British thermal units (mmBtu).	
Pricing Quotation	U.S. dollars and cents per mmBtu.	
Minimum Price Increment	$0.001 per MMBtu	
Maximum Daily Price Fluctuation	Initial Price Fluctuation Limits for All Contract Months. At the commencement of each trading day, there shall be price fluctuation limits in effect for each contract month of this futures contract of $1.50 per MMBtu above or below the previous day's settlement price for such contract month. If a market for any of the first three (3) contract months is bid or offered at the upper or lower price fluctuation limit, as applicable, on Globex it will be considered a Triggering Event which will halt trading for a five (5) minute period in all contract months of the NG futures contract, as well as all contract months in all products cited in the Associated Product Appendix of rule 220.08. Trading in any option related to this contract or in an option contract related to any products cited in the Associated Product Appendix which may be available for trading on either Globex or on the Trading Floor shall additionally be subject to a coordinated trading halt.	

Termination of Trading	Trading of any delivery month shall cease three (3) business days prior to the first day of the delivery month. In the event that the official Exchange holiday schedule changes subsequent to the listing of a Natural Gas futures, the originally listed expiration date shall remain in effect. In the event that the originally listed expiration day is declared a holiday, expiration will move to the business day immediately prior.
Listed Contracts	CME Globex: 118 consecutive months
	CME ClearPort and Open Outcry: The current year plus the next 12 years
Settlement Type	Physical

Source: CME Group, "Henry Hub Natural Gas Futures Specs" (2019).

Unleaded Gasoline (RB) – NY Harbor

Product Symbol	RB	
Venue	CME Globex, CME ClearPort, Open Outcry (New York)	
Hours (All Times are New York Time/ET)	CME Globex:	Sunday – Friday 6:00 p.m. – 5:15 p.m. (5:00 p.m. – 4:15 p.m. Chicago Time/CT) with a 45-minute break each day beginning at 5:15 p.m. (4:15 p.m. CT)
	CME ClearPort:	Sunday – Friday 6:00 p.m. – 5:15 p.m. (5:00 p.m. – 4:15 p.m. Chicago Time/CT) with a 45-minute break each day beginning at 5:15 p.m. (4:15 p.m. CT)
	Open Outcry:	Monday – Friday 9:00 AM to 2:30 PM (8:00 AM to 1:30 PM CT)
Contract Unit	42,000 gallons	
Price Quotation	U.S. dollars and cents per gallon.	
Minimum Fluctuation	$0.0001 per gallon	
Maximum Daily Price Fluctuation	Initial Price Fluctuation Limits for All Contract Months. At the commencement of each trading day, there shall be price fluctuation limits in effect for each contract month of this futures contract of $0.25 per gallon above or below the previous day's settlement price for such contract month. If a market for any of the first three (3) contract months is bid or offered at the upper or lower price fluctuation limit, as applicable, on CME Globex it will be considered a Triggering Event which will halt trading for a five (5) minute period in all contract months of the RB futures contract, as well as all contract months in all products cited in the Associated Products Appendix of rule 191.07. Trading in any option related to this contract or in an option contract related to any products cited in the Associated Products Appendix which may be available for trading on either CME Globex or on the Trading Floor shall additionally be subject to a coordinated trading halt.	
Termination of Trading	Trading in a current delivery month shall cease on the last business day of the month preceding the delivery month.	

Listed Contracts	36 consecutive months
Settlement Type	Physical

Source: CME Group, "RBOB Physical Futures Contract Specs" (2019).

Heating Oil (HO) – NY Harbor

Product Symbol	HO	
Venue	CME Globex, CME ClearPort, Open Outcry (New York)	
Hours (All Times are New York Time/ET)	CME Globex:	Sunday – Friday 6:00 p.m. – 5:15 p.m. (5:00 p.m. – 4:15 p.m. Chicago Time/CT) with a 45-minute break each day beginning at 5:15 p.m. (4:15 p.m. CT)
	CME ClearPort:	Sunday – Friday 6:00 p.m. – 5:15 p.m. (5:00 p.m. – 4:15 p.m. Chicago Time/CT) with a 45-minute break each day beginning at 5:15 p.m. (4:15 p.m. CT)
	Open Outcry:	Monday – Friday 9:00 AM to 2:30 PM (8:00 AM to 1:30 PM CT)
Contract Unit	42,000 gallons	
Price Quotation	U.S. dollars and cents per gallon	
Minimum Fluctuation	$0.0001 per gallon	
Maximum Daily Price Fluctuation	Initial Price Fluctuation Limits for All Contract Months. At the commencement of each trading day, there shall be price fluctuation limits in effect for each contract month of this futures contract of $0.25 per gallon above or below the previous day's settlement price for such contract month. If a market for any of the first three (3) contract months is bid or offered at the upper or lower price fluctuation limit, as applicable, on CME Globex it will be considered a Triggering Event which will halt trading for a five (5) minute period in all contract months of the HO futures contract, as well as all contract months in all products cited in the Associated Products Appendix of rule 150.07. Trading in any option related to this contract or in an option contract related to any products cited in the Associated Products Appendix which may be available for trading on either CME Globex or on the Trading Floor shall additionally be subject to a coordinated trading halt.	
Termination of Trading	Trading in a current month shall cease on the last business day of the month preceding the delivery month.	
Listed Contracts	Current Year + 3 Years + 1 Month	
Settlement Type	Physical	

Source: CME Group, "NY Harbor ULSD Futures Contract Specs" (2019).

WTI Crude – Houston (HCL) – Houston Ship Channel

Contact Unit	1,000 barrels
Price Quotation	U.S. dollars and cents per barrel
Trading Hours	Sunday - Friday 5:00 p.m. – 4:00 p.m. (6:00 p.m. – 5:00 p.m. ET) with a 60-minute break each day beginning 4:00 p.m. (5:00 p.m. ET)
Minimum Price Fluctuation	0.01 per barrel = $10.00
Product Code	CME Globex: HCL CME ClearPort: HCL Clearing: HCL
Listed Contracts	Monthly contracts listed for the current year and the next 3 calendar years. List monthly contracts for a new calendar year following the termination of trading in the December contract of the currennt year.
Settlement Method	Deliverable
Termination Of Trading	Trading terminates 3 business days prior to the twenty fifth calendar day of the month prior to the contract month. If the twenty-fifth calendar day is not a business day, trading terminates 3 business days prior to the business day preceding the twenty-fifth calendar day of the month prior to the contract month.
Position Limits	NYMEX Position Limits
Exchange Rulebook	NYMEX 201
Block Minimum	Block Minimum Thresholds
Vendor Codes	Quote Vendor Symbols Listing
Delivery Procedure	Delivery shall be made free-on-board ("F.O.B.") at Enterprise Crude Pipeline LLC's ("ECPL") ECHO crude oil terminal ("ECHO") or ECPL's Genoa Junction crude oil terminal ("Genoa Junction") or Enterprise Houston Ship Channel LLC's marine terminal ("EHSC") terminal in the Houston, Texas area. Delivery shall be made in accordance with all applicable Federal executive orders and all applicable Federa, State and local laws and regulations. The seller shall provide crude oil which is free from all liens, encumbrances, unpaid taxes, fees and other charges. For the pusposes of this rule, the term F.O.B. shall mean a delivery in which the seller. (1) provides WTI type light sweet crude oil at ECHO, Genoa Junction or EHSC at the point of connection between seller's incoming and buyer's outgoing pipeline or storage facility: and (2) seller retains title to, and bears the risk of loss for the crude oil to the point of delivery. Delivery at ECHO will be delivered at par to the final settlement price. Buyers and sellers that nominate to take or make delivery at EHSC or Genoa Junction shall be subject to a fee as administered by Enterprise.

Grade And Quality	At buyer's option, the buyer shall nominate one of three Enterprise facilities for receipt of crude oil and such delivery shall be made by any of the following method: (1) by interfacility transfer 7 ("pumpover") into one of the three designated Enterprise facilities; (2) by in-line (or in-system) transfer, or book-out of title to the buyer; or (3) if the seller agrees to such transfer and if the facility used by the seller allows for such transfer,without physical movement of product, by in-tank transfer of title to the buyer. Buyer retains title to, and bears the risk of loss for the crude oil at and from the point of delivery. Buyers and sellers that take or make delivery at ECHO will b subject to no additional service fee by Enterprise. Buyers that take delivery at EHSC or Genoa Junction will be subject to an additional service fee by Enterprise, payable by the buyer. Sellers that make delivery at EHSC or Genoa Junction will be subject to an additional service fee by Enterprise, payable by the seller. If both buyer and seller take and make deliveries at EHSC or Genoa Junction, both buyer and seller will be subject to an additional service fee by Enterprise, payable by Enterprise, payable each by both buyer and seller.
	Please see rulebook chapter 201

Source: CME Group, "WTI-Houston Crude Oil Futures Contract Specs" (2019).
Note: NYMEX symbol month codes are as follows:

F = January   **K** = May   **U** = September

G = February   **M** = June   **V** = October

H = March   **N** = July   **X** = November

J = April   **Q** = August   **Z** = December

Settlement Procedures

For the "near" or, "prompt" month, defined by NYMEX as the "active" month, the Daily Settlement price is the "Volume Weighted Average Price" (VWAP)of all trades that occur in the last (2) minutes of trading (14:28 pm to 14:30 pm, US Eastern Time Zone). This is the price one sees reported daily as the price of oil. For instance, a publication may say, "Oil settled at _____ today."

On the last day of trading for a futures contract (expiration), the "Final Settlement" is calculated as the VWAP of all trades which occur during the last (30) minutes of trading (14:00 pm to 14:30 pm, US Eastern Time Zone.)[5] This becomes the historically-recorded price for that month's futures price. It is the price against which all outstanding derivative positions tied to that contract are settled.

Summary Points

- NYMEX contracts are legally binding obligations to buy or sell energy commodities.
- Contracts are standardized.
- Each commodity contract has volume, price, location, and date specifications.
- Futures contracts provide price discovery.
- The NYMEX trades five energy commodities along with two precious metals.
- Trading occurs both in the pits of the exchange (open outcry) and via electronic platforms.
- The exchange has a unique set of symbols to identify the commodity/month/year.
- Market participants include commercial entities, or those interested in the physical commodity, and noncommercial traders (speculators), or those purely interested in profit. The latter category includes arbitrageurs.

Review Exercises

1. What is a futures contract?
2. What is the trading in the pits of the NYMEX called?
3. Name the NYMEX symbols for crude, natural gas, unleaded gasoline, and heating oil.
4. What is the contract delivery point for crude oil?
5. What is the contract delivery point for natural gas?
6. What is the contract delivery point for unleaded gasoline?
7. How is the monthly final settlement price determined for NYMEX commodities?
8. What is the contract delivery point for heating oil?
9. How is the daily settlement price determined for NYMEX commodities?
10. What are the days and hours of trading on the NYMEX (including electronic trading)?

Notes

1. Steven Errera and Stewart L. Brown, *Fundamentals of Trading Energy Futures and Options*, 2nd ed. (Tulsa: PennWell, 2002), 5.
2. Hull, John C., *Options, Futures, and other Derivatives*, 9th ed. (London: Pearson Education)
3. https://www.cmegroup.com/stories/index.html
4. In October 1973, OPEC embargoed sales of crude oil to Western countries that had supported Israel during the Arab-Israeli War. The embargo lasted about six months.
5. https://www.cmegroup.com/confluence/display/EPICSANDBOX/NYMEX+Crude+Oil

5

Mechanics of Futures Markets

Key Learning Points

- NYMEX futures contracts are legally binding obligations with the potential for both legal and civil consequences for nonperformance.
- Executing NYMEX futures contracts is done in a purposeful manner to support the binding agreement.
- Only licensed brokers and clearing parties can execute NYMEX futures contracts in the pits of the exchange or on its Globex electronic trading platform.
- There are three main groups of traders on the NYMEX: hedgers, traders, and arbitrageurs.
- Each commodity traded on the NYMEX has its own unique symbol, which includes the commodity type, month, and year.
- There are several market order types that can be placed, depending on the needs of the party wishing to trade.
- Futures can be executed in multi-month increments known as *strips*.
- Margin funds are required to be deposited to insure against potential losses.

The negotiation of NYMEX futures takes place either in the traditional trading pits on the floor of the Exchange or, via electronic trading platforms. Pit trading is also known as *open outcry trading* in that traders literally yell out their buy/sell orders and communicate with complicated hand gestures.[1] Electronic trading platforms such as NYMEX's Globex, however, have largely replaced pit trading. In either case, trades can only be executed by licensed brokers who undergo extensive training regarding the regulations pertaining to commodity futures, as well as the policies and procedures of the NYMEX. In addition, a "clearing" member must be used to guarantee the transaction.

The actual initiation and execution of a futures transaction has its own nuances that must be adhered to so that both parties are in agreement with the specific terms of the trade. Once a client gives a broker an official order, the client is bound by the outcome of the execution of the futures position. This is why phone calls are routinely taped so that any discrepancy in the execution of the orders can be reviewed, if needed. "Let's check the tapes" is a frequent solution to any misunderstandings about orders placed and filled.

There are three main groups of traders: hedgers, speculators, and arbitrageurs. *Hedgers* execute trades on behalf of their clients. These clients are using futures contracts to reduce their price and supply/market risk and pay the broker a commission. *Speculators* take positions for pure profit either for their own accounts or on behalf of their companies, which can include financial institutions of all types. *Arbitrageurs* look for price differences known as *spreads*, which represent an opportunity to make money based on price variances between related commodities (crude/gasoline), different time frames (May vs. December natural gas), and different locations (WTI vs. Brent). (Spreads will be covered in a later chapter.)

NYMEX Futures Quotes

NYMEX energy commodity prices can be accessed from a variety of sources both by subscription, such as Bloomberg's, and by publicly available media, including the NYMEX website.[2] NYMEX prices are also reported on other financial websites and are included in daily business publications such as the *Wall Street Journal*. The transfer of price data from the exchange to the various media outlets performs the price discovery function that is a hallmark of public exchange trading.

When viewing prices on a quote screen, the format is relatively consistent. Figure 5–1 shows a sample screen with NYMEX crude oil quotes.

Column labels

The column labels commonly appear in quotation marks and are defined in the following discussion.

"Symbol." This will be the NYMEX code for the particular commodity, month, and year, which differs depending on whether the trades are occurring during the regular pit session or on the Globex electronic trading platform. Common NYMEX symbols for the trading months (regular session) are as follows:

Crude = CL; Natural Gas = NG; Unleaded Gasoline = RB; Heating Oil = HO.

F = January	K = May	U = September
G = February	M = June	V = October
H = March	N = July	X = November
J = April	Q = August	Z = December

So for February 2019 WTI, the symbol is CLG9.

Description	Last	Net Change	Bid	Ask	Open	High	Low	Close	Volume
Feb 19 NYMEX Light Sweet Crude Oil (WTI) Futures Electronic	52.34	0.23	52.33	52.35	52.00	52.52	51.26	52.11	647599
Mar 19 NYMEX Light Sweet Crude Oil (WTI) Futures Electronic	52.64	0.25	52.63	52.64	52.29	52.81	51.55	52.39	243401
Apr 19 NYMEX Light Sweet Crude Oil (WTI) Futures Electronic	52.98	0.28	52.95	52.97	52.60	53.10	51.90	52.70	58569
May 19 NYMEX Light Sweet Crude Oil (WTI) Futures Electronic	53.33	0.25	53.35	53.37	52.79	53.44	52.30	53.08	42370
Jun 19 NYMEX Light Sweet Crude Oil (WTI) Futures Electronic	53.80	0.33	53.78	53.80	53.30	53.86	52.69	53.47	51012
Jul 19 NYMEX Light Sweet Crude Oil (WTI) Futures Electronic	54.09	0.31	54.13	54.15		54.18	53.05	53.78	20981
Aug 19 NYMEX Light Sweet Crude Oil (WTI) Futures Electronic	54.33	0.32	54.37	54.40	53.80	54.37	53.27	54.01	9584
Sep 19 NYMEX Light Sweet Crude Oil (WTI) Futures Electronic	54.50	0.33	54.54	54.56	54.18	54.62	53.38	54.17	12021
Oct 19 NYMEX Light Sweet Crude Oil (WTI) Futures Electronic	54.60	0.33	54.65	54.67		54.73	53.63	54.27	2687
Nov 19 NYMEX Light Sweet Crude Oil (WTI) Futures Electronic	54.66	0.34	54.71	54.74	54.36	54.69	53.61	54.32	2364
Dec 19 NYMEX Light Sweet Crude Oil (WTI) Futures Electronic	54.77	0.42	54.74	54.77		54.82	53.57	54.35	28895
Jan 20 NYMEX Light Sweet Crude Oil (WTI) Futures Electronic	54.73	0.39	54.74	54.78	54.40	54.73	53.73	54.34	896

Fig. 5–1. NYMEX WTI Crude Oil Futures Quote Screen

"**Last.**" The price of the last trade that occurred.
"**Bid.**" The current price a Buyer is willing to pay.
"**Ask.**" The current price a Seller is willing to accept.
"**Net Change.**" The last trade's price compared with the prior day's settlement price.
"**Open.**" The first trade of the day that occurs once the session opens.
"**High.**" The highest price traded so far that day.
"**Low.**" The lowest price traded so far that day.
"**Close.**" The prior day's settlement price.
"**Volume.**" The number of contracts traded so far that day.

Types of Orders

There are various types of orders that can be placed on the exchange depending on the method selected by the customer.

Market order. The most common type of order used. With this type of order, the client has asked to have the order filled immediately at the current market price, regardless of where it is, since the customer usually has obtained knowledge as to where the market is trading.

Limit order. The client has specified a price at which the order can be executed, but the trade cannot be executed if the price is not favorable for the client. For instance, a buyer of crude places a limit order at $50.00/bbl. The trader for the client can only buy at $50.00/bbl. or below. If the market stays above $50.00/bbl, the order is not filled.

Stop order (or "stop-loss" order). The purpose of this type of order is to limit losses on a position opened when the order is initiated. For instance, a trader may want to sell crude oil at $50/bbl but believes prices are going to increase. The trader wants to be able to sell later at a higher price. If, however, prices should fall, the trader wants to minimize potential losses. So at the time the buy order is placed, the trader specifies a $45/bbl *stop*, meaning that the position is closed out immediately if the market falls to $45/bbl. Each trader must determine his or her appetite for risk when setting the stop price level. Keep in mind that prices can fluctuate dramatically throughout the day, so if the trader had placed a stop at $48/bbl, it could easily have been reached, and the position would have been closed automatically. Setting the lower $45/bbl stop, however, gave the trader the chance to withstand any downward movement that might reverse itself during the trading session.

Market-if-touched. This order is used if traders have open positions and wish to get a specific price at which they will make a profit. For example, a trader could be long crude contracts at $50.00/bbl and wants those to be sold if the market hits $51.00/bbl. In this case, if a bid occurs at $51.00/bbl (the price is *touched* in the market), the trader's order is filled, and the position is closed. Again, should prices not reach the trader's requested price, nothing occurs.

Some types of orders have time restrictions, such as the following:

Fill-or-kill. Clients have given the broker a specific price or other parameters. If the order cannot be filled immediately, it is withdrawn.

Open (or good-'til-canceled). When this type of order is used, the client has placed a trade such that if the market does not reach the desired price level, the order remains open until the client cancels it or the contract reaches expiration, whichever occurs first. If the current market price for crude is $50.00/bbl, a client wishing to go long at $45.00/bbl may place a buy order at $45.00/bbl, "GTC" (good 'til canceled). That order remains in effect until it is filled, the client cancels it, or the contract expires. When placing this type of order, the client must watch the market carefully and remember the order. Days, weeks, or months could go by without the price level having been achieved. Then the order could suddenly be executed if the specified price trades. By then, this price level may be counter to the client's current goals or position. Often the trader has picked a price with a slim chance of occurring, hoping to make substantial gains if it does happen.

With the exception of "open" and "GTC" orders, most orders are deemed to be for the current trading day only and expire upon the end of that day's session unless the client gives instructions otherwise.

Strips

Futures contracts can be bought or sold in multimonth trades known as *strips*. Any order for futures that is two months or longer is considered a strip. They are a convenient way of executing trades in that the average price for the period is used as opposed to individual month orders. Perhaps a trader wishes to sell one crude oil futures contract per day for an entire year. Rather than the order being executed for each and every month, the exchange will provide an average price for the 12 months. The following are some common strip periods:

- 90-day—first 3 futures months
- 6-month—first 6 futures months
- 12-month—starts with the "prompt" or "near" month and goes out for 12 consecutive months
- Calendar ("Cal")—the entire period of a calendar year, e.g., January through December of the same year. It does not start with the prompt month unless that month happens to be January.
- Summer—April through October of a given year
- Winter—November through March
- Core winter—December through February

Recall that "summer" and "winter" in energy trading are not comprised of the months normally associated with the four seasons. Historically, energy demand has been higher in the winter than in the summer. As such, utilities store energy (mostly heating oil and natural gas) from April through October during periods of lower demand and cheaper pricing to prepare for the colder weather months of November through March. While there is no guarantee that prices will be lower in the warmer months than the colder ones, this early pattern of buying and storing has established the seasonality of the energy futures markets.

Margins

Once an order is placed and executed, the investor (trader, hedger, etc.) will be required to post funds as credit assurance against any potential losses that position could face. (Remember that the exchanges and some OTC markets require trades to be cleared through a financially stable entity. Those clearinghouses, in turn, demand credit from their clients.) These funds are placed in a margin account, and this first amount is designated as the *initial margin*. As mentioned earlier, the exchanges calculate daily settlement prices for each commodity and for each month for which contracts traded that day. Those closing prices can change the value of the open positions investors have when they are compared to the settlements, a process known as *mark to market*.

For example, an investor buys five January WTI contracts at $61.20/bbl and posts the required margin per contract. If the daily settlement price on the first day is $61.10/bbl, the investor would be selling the contracts at a loss of $0.10/bbl, or −$500 (5 contracts × 1,000 bbl × $0.10). Now, the investor has not yet sold the contracts. If the investor were to do so, however, based upon the known market price at the time, he or she would lose $500. As long as the positions have not been closed, they are subject to price volatility. Any open position for which an adverse price movement occurs essentially means that position is "losing" money, although it is an unrealized loss. Nonetheless, on the books, that position would result in a loss should it be closed out immediately. Therefore, additional funds, or a *maintenance margin*, must be placed in the margin account to offset the negative price movements. We previously mentioned that the nature of financial derivatives is a zero-sum game; for every winner, there is a loser. So it is with margins. Those whose positions become negative will have to supplement their accounts, while those whose positions increase in value will see funds added.

Summary Points

1. NYMEX futures contracts are legally binding obligations to make or take delivery of the underlying physical energy commodity.
2. Only licensed brokers can execute NYMEX contracts.
3. Several order types can be entered into.
4. NYMEX market participants include Hedgers, Traders and, Arbitrageurs.
5. Margin funds are required when placing NYMEX orders to protect against potential losses.

Review Exercises

1. What is the NYMEX electronic trading platform?
2. What does CLH9 refer to?
3. What are the three main types of traders on the NYMEX?
4. The summer strip covers what months?
5. The winter strip covers what months?
6. Name five types of NYMEX orders that can be placed.
7. What is a calendar strip?
8. What are the funds deposited in an account to offset the potential losses of a *new* open position known as?
9. What are the funds deposited in an account to offset potential losses of an *existing* open position known as?

Notes

1. On February 4, 2015, CME Group announced the phase-out of futures pit trading. See CME Group, "CME Group to Close Most Open Outcry Futures Trading in Chicago and New York by July; Most Options Markets to Remain Open" (February 4, 2015), http://www.cmegroup.com/media-room/press-releases/2015/2/04/cme_group_to_closemostopenoutcryfuturestradinginchicagoandnewyor.html.
2. https://www.cmegroup.com/market-data/delayed-quotes/energy.html

6

Using NYMEX Contracts for Trading and Hedging

Key Learning Points: NYMEX Trading

- Trading of energy financial derivatives for profit is known as *speculation*.
- Generally, traders are comprised of four groups: scalpers/day traders, arbitrageurs, "pure" day traders, and position traders.
- Speculative trading adds liquidity and efficiency to the market.
- High-frequency trading is speculative trading using supercomputers running complex mathematical algorithms, resulting in extremely high trading volumes.
- The Commitments of Traders reports reveal positions held by commercial and noncommercial market participants.

Trading of NYMEX Futures Contracts

Trading of energy financial derivatives for pure profit (speculation) generally involves the fundamental research delineated earlier, as well as the analysis of technical charts, with the latter being the primary tool of the day trader/investor.[1] "Buy Low/Sell High" is the trader's credo. Of course, "Sell High/Buy Low" can also reap rewards if one has a more bearish outlook.

Speculators take many forms:
- *Scalpers.* Scalpers are mostly day traders who trade for their own book. They can be large firms or locals, and they add to market liquidity. Scalpers are often *market makers.* They are willing to assume the risk of those needing to lay off risk, thereby giving the market makers a position to trade.
- *Arbitrageurs.* Arbitrageurs are spread traders looking at absolute price relationships between (and among) energy derivatives for profit opportunity.

- *"Pure" day traders.* These are speculative traders looking to make a profit in very short-term day trading. They do not normally hold positions beyond a day and do not "make market."
- *Position traders.* These traders take positions and hold them for longer periods.
- *Hedging.* This is any action taken to reduce one's risk.
- *Commercial entities.* Commercial entities use hedging to reduce their price and/or market risk.

Regardless of their market perspective, the goal of speculators is to enter and exit the market profitably. Traders or investors will take a position once they determine what direction they believe prices are heading. If they are *bullish* on prices, their perspective is that prices are going to rise. So in order to make money, they will buy now and sell when prices move higher, which they think will occur. Conversely, if they are *bearish* on prices, their view is that prices are going to fall. In order to gain a profit, they will sell now and buy when prices retreat, which is the price direction their analysis indicated. However, neither position is guaranteed to be successful.

All traders and investors draw their own conclusions as to market direction and act accordingly. Again, there can be no buyer without a seller, and vice versa. Differing views on market price direction serve to sustain the liquidity in financial energy derivative markets.

Another avenue for the trading of energy futures contracts for profit occurs when financial institutions execute trades on behalf of their clients and choose to assume the risk from the clients. That is, if a client places an order, the financial institution does have the option to transfer that risk immediately to the marketplace or to hold that position if analysis indicates there could be a favorable price movement.

For example, an oil producer wishes to sell some January WTI at the current market price of $61.20/bbl. When the producer contacts a financial counterparty to execute this trade, active market prices are trading above and below this level (the *bid/ask spread*). A financial counterparty that is bullish on prices may tell the client that the order has been filled at $61.20/bbl but not immediately sell WTI contracts in the market. The counterparty's hope is to gain not only the brokerage fee, but also a profit on the position should prices soon exceed $61.20/bbl. This is known as "taking it into the book" held by the financial counterparty. The *book* consists of the counterparty's total open proprietary trading positions. Either way, the producer receives the desired sell price, which is guaranteed by the financial counterparty regardless of whether or not the counterparty decides to assume the price risk.

Speculative trading adds to both the liquidity and efficiency of the futures marketplace, with "high frequency trading" (HFT) perhaps being the biggest contributor to the volume of contracts executed these days. Supercomputers programmed with complex algorithms are used to place orders for thousands and thousands of contracts in nanoseconds. The order triggers are merely reactions to price direction movements and do not involve consideration of fundamental or technical information. While some may argue that computers are dominating the market and setting prices, the huge volumes executed by these programs provide opportunities for various parties to find counterparties to the trades they need in the market. Thus the market has actually become more efficient through the enhanced liquidity brought about by HFT.[2] (This is also referred to as *algo-trading*.)

The Commitments of Traders Reports: A Measure of Speculation and Hedging

The US Commodity Futures Trading Commission, the federal regulatory body with jurisdiction over energy futures markets, monitors the open positions (those not yet settled) on NYMEX. Every Tuesday, the CFTC issues a report showing a breakdown of open positions by market participant type, commercial and non-commercial. Commercial entities are industry participants using the market for hedging, while non-commercial entities represent the pure speculative positions. This piece of market intelligence can signal much about the sentiment of the players. Large speculative shorts would imply that traders think prices are going lower, while large long positions would imply the opposite. For commercial interests, short futures positions would be a sign that producers are hedging, since they are selling forward. Long put option volumes are another signal that producers are hedging. Long futures and call options positions would indicate buying activity by commodity end users.

Risk Management and Hedging Using NYMEX Futures Contracts Key Learning Points: Energy Risk Hedging

- Producers and consumers of energy can reduce both their physical (market or supply) and price risk using financial derivatives such as futures and forwards.
- Futures are exchange-traded contracts backed by the underlying commodity. On NYMEX, these are the energy commodities, including crude oil, natural gas, heating oil, and unleaded gasoline.
- Forwards are over-the-counter financial instruments that can be used to hedge price risk.

- Hedgers are not speculators.
- In a hedge, commercial participants take the opposite position from their physical position in the financial markets.
- Financial positions must be settled monthly.
- Simple risk reduction can occur through the use of basic financial derivatives:
    - NYMEX futures contracts
    - Forwards
- Hedging reduces both physical and financial risk.
- Hedging is performed by commercial entities. It is not "trading."
- The concepts of *parallelism* and *convergence* are key to the successful outcome of any hedge.
- Hedgers have two positions, one financial and one physical.
- Hedgers must take an opposite position in the financial market to the one they have in the physical market.
- Commodity producers are long the physicals and must sell the financials.
- Commodity consumers are short the physicals and must buy the financials.
- Multimonth and multiyear hedges, or strips, can be obtained.

Hedging

A firm that actively uses derivatives and other techniques to alter its risks and protect its profitability is engaging in **risk management**.

—Robert L. McDonald

Risk management is the process of identification, analysis, and either acceptance or mitigation of uncertainty in investment decision making. In addition, risk management may be viewed as immunization of a portfolio against risk.

—Iris M. Mack

Before we begin our discussion of how energy companies can use financial commodity instruments to hedge their risk, we must define the risks these companies face and specify which of those can be mitigated using the financial

energy derivatives markets. Dahl cites (from Jorion) five types of general financial risks facing energy companies:[3]
- *Market risk* relates to price changes of financial assets and liabilities.
- *Credit risk* relates to defaults on contractual obligations.
- *Liquidity risk* relates to the lack of market activity or to a failure to meet a cash flow obligation.
- *Operational risk* relates to technical problems with financial trading systems and fraud.
- *Legal risk* relates to losses from failure to comply with the law or from adverse regulatory changes.

Since we are going to address the commercial participants who enter the financial commodities arena purely for risk reduction, some of these risks need to be further refined, and other risk exposures need to be addressed.
- *Market risk.* For energy companies, this really encompasses three different risks:
  - Price risk. Price risk is perhaps the most important risk for producers and consumers of energy commodities. The price received for the goods or services produced drives the business.
  - Market risk. Producers of energy commodities seek a guaranteed market for future production.
  - Supply risk. End users of energy commodities need a guaranteed source of supply for future needs.
- *Credit risk.* Commercial entities must have a credit rating sufficient to satisfy the financial counterparty involved in the trade execution. On the other hand, insolvency of the financial institution counterparty is also a possibility. Given the events of the past decade, it is apparent that credit risk is truly a bilateral risk.
- *Liquidity risk.* In order to satisfy one's desired hedging goals, a sufficient number of counterparties must be available in the marketplace for the volumes and time frames needed, and at a price based on adequate competition. Generally speaking, trading periods beyond three years lack a substantial number of market participants.
- *Operational risk.* For commercial entities, this really pertains more to their actual ability to satisfy the requirement to "make or take" the commodity. Once committed, a disruption in supply obligated to be delivered by a producer or the inability of an end user to consume committed volumes can result in default under the financial agreements. This can result in both financial and legal penalties.[4]
- *Legal risk.* In addition to those stated above, "contractual" risk faces energy companies in both the provisions of the International Swaps and Derivatives Association (ISDA)[5] agreements and the bilateral physical

commodity contracts. "Special provisions" can be added as amendments to these agreements that can alter the meaning and intent of the original base contract.

- *Execution risk.* While liquidity risk covers the exposure to a lack of counterparties, execution risk pertains to the ability to get an order filled as desired. When using third parties, with or without electronic platforms, there is always a lag between placing an order and the execution of that order. Once an order is placed, the intermediary has an opportunity to fill it somewhere near the requested price, but not necessarily at it, due to the ever-changing prices. When the client's price can be filled, but the third party actually executes a price that is better, the additional margin gained is known as *slippage*. For example, a producer wants to sell crude futures at $61.20/bbl and places that order. If the market when the broker goes to execute this sale is $61.205/bbl, the broker can tell the producer the order was filled, and the broker pockets the extra $0.005/bbl. While this may not seem like a lot, on an order of just one oil contract per day for one month, the broker would make an extra $150. This is in addition to the brokerage fee charged to the producer, and the broker can do this several hundred times a day and for longer-term trades. More slippage can be earned in the normal pit trading environment due to the longer execution time, which is why many traditional floor traders do not want to see that manner of trading end.

In the previous chapters, we learned that NYMEX energy contracts represent the actual *right to buy or sell energy commodities*. So for the commercial market participants, these provide both a market for production and a source of supply, at a fixed price. For instance, producers of natural gas, crude oil, or refined products (heating oil and gasoline) can sell financial contracts, thus guaranteeing that they will have a firm market in the future at a fixed price. Conversely, consumers of these same products can buy contracts to ensure that they will have a firm supply source in the future at a set price. Utilizing financial contracts to reduce price and/or commodity risk is known as *hedging*.

Mack presents some key reasons why market participants may enter into energy derivative contracts:[6]

- Reduce exposure to price risk by shifting that risk to market participants with opposite risk profiles[7] or participants who are willing to accept the risk in exchange for profit opportunity.[8]
- Lock in prices and margins.
- Minimize potential for unanticipated loss.
- Provide arbitrage opportunities.
- Improve credit worthiness.[9]
- Increase borrowing capacity.[10]
- Augment financial management and performance capabilities.

In this chapter, we will focus the ways in which commercial players in the energy industry use the financial markets for hedging their price and market risks.

In chapter 3, we defined energy futures contracts and explained the function of the NYMEX. We also identified the two main participants in financial energy markets as commercial and noncommercial.

Commercial entities have an interest in the commodity itself due to the particular business they are engaged in. For example, an oil refiner not only needs actual physical crude oil as feedstock but also has an interest in its future price. Since the feedstock is necessary for all of the refined products they produce, their profitability is impacted by the purchase price of crude.

In addition, refiners sell products such as gasoline and heating oil, both of which are traded in the financial markets. So the refiner's gross profit, or spread, can be said to be dependent on the feedstock price for crude (cost) and the market price for what it produces (revenue).

On the other hand, exploration and production companies need to know the future market price for the crude oil they will extract from their wells. This holds true not only for the revenue they can derive through future sales of production but also for valuation of their reserves, a critical component of the collateral they use to borrow for drilling and production capital. (*Note:* Reserve values are normally re-determined biannually. Dramatic decreases in those values can trigger default provisions within banking covenants or investor agreements tied to the lending. This situation was very prevalent during the sustained fall in oil prices from 2014 through 2016. E&P companies, already reeling from lower prices and high debt, faced the additional burden of the devaluation of their collateral, making it that much harder to sustain ongoing operations, let alone future drilling and production activities.)

The same situation holds true for natural gas. Natural gas is a component of manufacturing costs in such industries as steel manufacturing, fertilizer production, and food processing. In the power industry, the price of natural gas impacts the cost of generating electricity. And for midstream processors, natural gas is the main component for the production of valuable natural gas liquids (NGLs).

As with oil, E&P companies that produce natural gas can also see the future market prices for their production. Keeping in mind that futures contracts are *legally binding obligations to buy or sell a commodity*, they guarantee a market for producers and a source of supply for consumers. They also guarantee a set or fixed price, thereby reducing price risk as well.

As explained by Mack, "A hedge is an investment that reduces the risk of adverse price movements in an asset." She adds that "hedging against risk involves the purchasing of financial instruments to offset the risk of adverse price movements."[11]

As previously explained, when commercial parties enter the financial energy marketplace to reduce their supply and/or price risk, it is known as hedging. This is much the same as one who bets on the "favorite" in a horse race but hedges that bet by also placing bets on other possible winners. The hope is to mitigate losses if the favored horse does not win. In energy commodities, the party hedging is concerned about market movements that could adversely affect their profitability. According to Mack, "If a market participant employs a hedge strategy it may incur an inevitable tradeoff between risk and return."[12] (As we will see, certain hedges, while protecting negative price impacts, also limit the positive impact of favorable gains.)

In order to hedge supply and price risk correctly, *physical players must take a financial position that is* **opposite** *to their physical position.* For instance, a producer has a commodity and needs a market. (They are said to be *long* the commodity.) In the futures market, they will sell contracts and thus create a future market for their natural gas, crude, etc. This guarantees that a counterparty will take the production and will do so at a known, fixed price and, at a set time and specified location.

Consumers need the energy source. (They are said to be *short* the commodity.) Therefore, they must buy contracts in the futures markets. For them, this guarantees that a counterparty will provide the commodity and will do so at a known, fixed price and at a set time and specified location.

In chapter 3, we mentioned that less than 2% of all futures contracts actually go to delivery; that is, the physical commodity does not usually change hands as a result of the financial transactions. (Think about the non-commercial players. They neither have, nor want, the actual physical commodities. They are just trading price.[13])

So how exactly does hedging work?

Futures prices for any commodity are deemed to represent "the market" as it is known at the moment. (We also addressed in chapter 3 the idea of the price discovery that futures markets provide.) Producers are considered to have sold "at market" at the time they enter into futures contracts. But we know that prices will change between the time this deal was transacted and the time the commodity actually changes hands. These fluctuations in the futures prices will impact the cash prices, and vice versa. When the actual delivery month arrives, the "real" physical prices are posted in the publications we referenced in chapter 2. As we stated, these are established through physical cash trading for that month.

Therefore, cash and futures prices influence each other throughout the trading of the *near month*.[14] This relationship leads them to move in-sync, a concept known as *parallelism*, and they approximate one another as they near the final settlement of the financial contracts, a concept known as *convergence*. Both principals are critical to the success of a hedge.[15]

Let's look at some examples of simple fixed-price hedges for producers and consumers of natural gas.

A producer of natural gas in the United States wishes to sell some December production in June at the current market levels, which will help the company meet established earnings targets. The producer also has concerns that future prices may be lower than current levels. To hedge the price risk that can occur between now and then, the producer will *sell* the December financial NYMEX futures contracts. This guarantees the producer a market at Henry Hub at a fixed price when the December production month comes around.

For example, let's say the December NYMEX contract is currently trading $3.10/MMBtu. The producer decides to execute the hedge. As mentioned above, the producer will *sell* December NYMEX contracts (the opposite of the producer's physical position). The producer's price is now set at $3.10/MMBtu for the sale of December Henry Hub natural gas. If prices should increase between the date of this transaction and December, the producer cannot capture the upside. But the producer chooses to forego this opportunity to guarantee obtaining a specific price. The producer is thus protected against lower prices if the market drops. (*Note:* Contrary to some opinions, the producer does not "lose" money if prices go higher than the hedged price. The only way that can happen is if the producer hedged at a price that was less than its known lifting costs. To be sure, if prices do go higher, there will be some 20/20 hindsight second-guessing. But the firm made a business decision that the prevailing December price represented an opportunity to fulfill a portion of its financial goals.)

At the end of November, when the December futures contract expires, the producer decides to close out its financial position. So the producer must now *buy back* the contracts in order to balance its financial position on NYMEX. The producer's position with the exchange is now *flat*.

So what happens to the price that the producer will receive when actually selling its natural gas in the December cash market? Since the futures pricing represents the market, the December cash market prices rose and fell as the contracts traded.

That means that both the price of the futures contracts that the producer sold, as well as the cash price (market), fluctuated throughout the life of the December contract trading. The producer had to buy back the futures contracts on the final settlement day. If the futures contract price had risen, the producer took a loss on the financial transaction. But what happened in the cash market? Since futures rose, so did the cash market, thus providing the producer a corresponding gain in the physical market.

Conversely, if futures prices had fallen by the final settlement day, the producer would have paid less when buying the futures contracts back and would have made a profit on the financial transaction. However, since the futures market declined, so did the cash market, thus lowering the actual price the producer received when the December natural gas production was sold in the physical market. Table 6–1 illustrates the steps in this process and the financial or physical gain/loss.

Table 6–1. Simple fixed-price hedge for natural gas

Natural Gas Hedge Mechanics

1) Producer sells December natural gas contracts in June @ $3.10/MMBtu

2) December contract expires in late November - Producer closes-out position (buys back)

3) Producer sells physical natural gas in December production month at published cash price.

Scenario 1

December NYMEX Natural Gas Contract settles @ $3.50

1) Producer has to buy back NYMEX contracts @ $3.50 for a loss of ($0.40) per MMBtu.

2) Cash market fell in conjunction with fall in NYMEX; Producer loses on physical sale

Producer						
December NYMEX Sale	Buy December NYMEX Settlement (1)	Financial Gain/Loss	Sell December Physical Market	December Physical Gain/Loss		Net Gain/Loss
$3.10	$3.50	($0.40)	$3.50	$0.40		$0.00

(1) Closing-out position on Last Day

Scenario 2

December NYMEX Natural Gas Contract settles @ $2.70

1) Producer has to buy back NYMEX contracts @ $2.70 for a gain of +$0.40 per MMBtu.

2) Cash market fell in conjunction with fall in NYMEX; Producer loses on physical sale

December NYMEX Sale	Buy December NYMEX Settlement (1)	Financial Gain/Loss	Sell December Physical Market	December Physical Gain/Loss	Net Gain/Loss
$3.10	$2.70	$0.40	$2.70	($0.40)	$0.00

(1) Closing-out position on Last Day

On the other side of this market transaction is a midstream natural gas company, which gathers and processes natural gas and is dependent on the spread between the price of natural gas, which is the feedstock, and the natural gas liquids (NGLs) the midstream company produces. If they are concerned about rising natural gas prices in December, they can buy NYMEX contracts at $3.10/MMBtu and thus be guaranteed supply at Henry Hub at a fixed price when the December production month comes around.

When the end of November arrives, the midstream company also has to close out its financial position and sells its contracts on the final settlement day. If the contract price has fallen, the company would take a loss on this financial transaction. But what happened in the cash market? Since futures fell, so did cash, thus providing a lower price in the physical market for the midstream company.

Conversely, if futures prices had risen by the final settlement day, the midstream company would have made a profit on the financial transaction when it sold the contracts back. However, since the futures market rose, so did the cash market, thus raising the actual price the midstream company paid when it bought natural gas in the physical market.

Based on the concept of convergence, mentioned above, the final settlement price for the December NYMEX natural gas contract would represent the cash market price for that month. In both of these scenarios, the gain or loss in the financial market is offset by a corresponding and opposite gain or loss in the physical cash market. When there is a 1:1 correlation between the financial and physical markets, it is referred to as a *perfect hedge*.

This process can be performed many times over by producers and consumers as desired. Thus suppliers and end users can establish a fixed price and ensure themselves a market or supply for energy commodities that are financially traded. And theoretically, they can do so for as many future months as the particular contract allows (depending on the number of market participants willing to trade that far out, i.e., the liquidity).

For accounting purposes, both the financial and physical transactions are kept separately. So in actuality, a gain or loss in the financial reporting would appear, and the corresponding gain or loss in the reported physical price would appear.

Keep in mind that, for the purposes of the examples given, the energy commodities are being physically delivered at their respective NYMEX contract delivery points. We will address how to figure pricing for locations other than the financial hubs in a later chapter. In addition, crude oil comes in many grades, and prices are adjusted relative to the WTI price to account for variances. Natural gas, on the other hand, must meet the standard quality specifications of the various transporting pipelines, so only locational differences from the Henry Hub have to be considered.[16]

These simple fixed-price hedges form the basic building blocks for the more complex financial derivative hedges.

Summary Points

1. Trading in financial energy derivatives for pure profit is known as speculation.
2. Speculative trading adds liquidity and efficiency to the market.
3. High-frequency trading relies on complex algorithms carried-out on super-computers.
4. Energy futures contracts can be used to reduce price and/or market risk. This is known as, "hedging".
5. Hedgers are not Speculators.
6. Hedgers take financial positions that are the opposite of their physical positions.
7. The concepts of *parallelism* and *convergence* allow hedging to be effective.
8. Futures and/or forwards can be used to achieve simple, fixed-price, hedging.

Review Exercises

1. The two main NYMEX market participants are _____ and _____ entities.
2. Utilizing financial derivatives to reduce one's price and supply/market risk is known as _____.
3. Producers of the commodity are said to be _____ the physical commodity and therefore must _____ futures contracts.
4. Consumers for the commodity are said to be _____ the physical commodity and therefore must _____ the futures contracts.
5. Commercial entities must take a financial position that is the _____ of their physical position in order for the hedge transaction to be successful.
6. The concept of _____ explains how cash and futures prices track one another.
7. The concept of _____ explains how cash and futures prices approach each other as the futures contract reaches expiration.
8. When cash and futures price changes occur on a 1:1 but opposite basis, it is referred to as a _____.
9. Fixed-price hedges can be accomplished using two types of financial derivatives, _____ and _____.
10. Trading futures contracts for pure profit is known as _____.

Notes

1. Mack refers to speculators who utilize both fundamental and technical analysis as "techno-fundamentalists." See Iris M. Mack, *Energy Trading and Risk Management: A Practical Approach to Hedging, Trading and Portfolio Diversification* (Singapore: John Wiley & Sons, 2014).
2. As with most things today, videos are available on YouTube that provide good explanations of HFT order execution.
3. Carol Dahl, *International Energy Markets: Understanding Pricing, Policies, and Profits*, 2nd ed. (Tulsa: PennWell, 2015), 465.
4. Commercial entities hedging their physical risk are normally cautioned not to enter into a volumetric commitment that cannot be met currently or in the future.
5. More information about the International Swaps and Derivatives Association is available at https://www.isda.org/.
6. Mack, *Energy Trading and Risk Management*.
7. Other hedgers.
8. Market-makers.
9. Hedgers take the downside risk out of the market and establish revenue streams which improve their credit standing.
10. For energy producers who borrow capital for activities such as exploration and production, establishing a profit margin through hedging reduces their exposure to price uncertainty and makes their ability to pay back loans more secure. Some financial institutions will insist on some level of hedging as a condition of lending.
11. Mack, *Energy Trading and Risk Management*.
12. Ibid.
13. On September 28, 2016, more than 1,000,000 November 2016 CL contracts traded on NYMEX on word that OPEC would meet to discuss cuts in production output. This was the equivalent of 1 billion bbls of crude oil, a volume that was hardly going to exchange hands at Cushing, OK two months later.
14. The first futures contract that can be traded is known as the *near, prompt*, or *front month*.
15. Steven Errera and Stewart L. Brown, *Fundamentals of Trading Energy Futures and Options*, 2nd ed. (Tulsa: PennWell, 2002).
16. Every transmission pipeline sets quality standards for the receipt of natural gas. These specifications are found within the "General Terms and Conditions" of the respective pipeline's tariff. For an example, see Natural Gas Pipeline Company of America: https://pipeline2.kindermorgan.com/Documents/PDFView.aspx?code=NGPL&fname=NGPL_EntireTariff.pdf&pdftag=ftsp

7

Financial Energy Derivatives: Swaps

In chapter 6, we focused on the futures markets and how simple hedges can be accomplished using exchange-traded contracts. In this chapter, we will address the over-the-counter, non-exchange-traded markets. Keep in mind that NYMEX contracts are referred to as *futures*.

Key Learning Points: Financial Energy Derivatives—Swaps

- *Swaps* are exchanges of payments between two parties. They are strictly financial. No physical exchange of the commodity takes place.
- One party to a swap transaction agrees to pay a current market price (*fixed*), while the other agrees to pay a price in the future (*floating*).
- Swaps are normally *fixed-for-floating*, whereby one price is the current market price (fixed) and the other price is the future settlement price (floating).
- Swaps are a simpler and less expensive way to hedge price risk.
- Non-exchange-traded financial derivatives are over-the-counter (OTC) transactions and can be traded via electronic trading platforms or over the phone with licensed brokers.
- Swaps trade in the OTC markets.

According to Errera and Brown, *swaps* represent "an exchange of payments between two parties."[1] Hull defines swaps as "an over-the-counter derivatives agreement between two companies to exchange cash flows in the future."[2] They are financially settled and no physical commodity is delivered or received by either party. Mack defines swaps as a "derivatives contract in which counterparties exchange an asset (or liability) for a similar asset (or liability)."[3] She also defines a financial swap as "a derivative contract in which counterparties exchange cash flows of one party's financial instrument for those of the other party's financial instrument."[4] They represent a substitute for the futures contracts *but rely on NYMEX pricing* to establish the financial arrangement for the swap contract. For energy trading and hedging purposes, these can also be referred to as a "plain vanilla energy swap," which is "a derivatives contract in which counterparties exchange a floating price for a fixed price over a specified period of time."[5] (In the case of natural gas swaps, they are actually referred to as "Henry Hub look-alikes.")

Similar to a NYMEX contract, the elements of a swap contract include the commodity, location, date, and price. However, one of the advantages of swaps is the ability to customize the volume to suit the hedger's needs. (As discussed earlier, NYMEX futures volumes are fixed at 1,000 bbl. for crude, 10,000 MMBtu for natural gas, and 42,000 gallons for gasoline and heating oil. But what if you want to hedge more or less than these amounts? Swaps allow you to do just that.)

When swaps are traded bilaterally, the two parties involved will execute a legal document known as a *confirmation*, which spells out the details of the transaction. These are part of the base ISDA *Master Agreement* for conducting financial trades. (The International Swaps and Derivatives Association [ISDA] is based in New York and provides standardization methods for the trading of financial derivatives.)

We use the phrase *fixed-for-floating swap* to signify the prices agreed to by both parties in the contract. The *fixed price* is always the *current market price*, which is known at the time the contract is executed. The actual exchange of payments will occur when the NYMEX futures contract settlement price is known and is then compared to the fixed price. We refer to the settlement price as the *floating* or *unknown* part of the swap, since it is not known until the contract's last trading day (settlement). It "floats" as the NYMEX contract trades daily until then. The difference between the two prices represents the amount of payment due one party or the other.

For example, in February, a crude refinery is concerned about oil prices possibly rising as the summer driving season approaches. The June NYMEX WTI price is trading at $60.50/bbl, and the refiner wishes to fix a price now. So the refiner buys a crude oil swap from a financial counterparty at $60.50/bbl. In this swap, the refiner's position is exactly the same as if they traded on NYMEX, that is, the refiner buys the swap since it is short in the physical market. But the refiner is buying *price* protection *only*, and not a commodity-backed contract. The financial institution has sold the swap, just as a NYMEX counterparty would do. Each party has now guaranteed the other a price that will be financially settled upon expiration of the June WTI contract.

In late May, the June NYMEX WTI contract reaches a final settlement price of $60.75/bbl, so the floating part of the swap has now been determined, which represents the financial counterparty's swap price. The difference between the refiner's $60.50/bbl and $60.75/bbl is the amount exchanged between the two based upon their respective buy/sell positions. Each has to close out their positions *with one another* and not an exchange. However, the mechanics of this type of hedge are the same as if the counterparties had used NYMEX contracts.

In this case, the refiner bought the $60.50/bbl swap and now sells it back to the counterparty at $60.75. The counterparty pays the refiner a settlement of $0.25/bbl. The refiner had a *financial* gain of $0.25/bbl, but as with a NYMEX fixed-price hedge, prices rose in the physical market to $60.75/bbl, making the refiner's actual physical crude purchase price $0.25/bbl higher. So the refiner has a zero net monetary position with this hedge, just as if the transaction took place on NYMEX. But only the price was exchanged and not the underlying commodity.

If, however, the NYMEX settlement was $60.25/bbl, the refiner would have to pay the counterparty the $0.25/bbl difference and take a loss on the financial transaction. Once again, if NYMEX went down $0.25/bbl, so did the cash market, which saved the refiner that amount on the physical crude purchase. Again, the refiner has a zero net monetary position with this hedge approach.

Table 7-1 shows the standard, fixed-price, hedge for crude oil using NYMEX futures contracts. The calculations for this Swap transaction are the same as those shown in table 7–1. But, in Table 7-2, we substitute the word, "Swap" for "NYMEX" to signify the change in financial instrument we are using. As you will note, the prices don't change. We are still relying on NYMEX for the pricing of both the "fixed" and, "floating" pieces of our Swap. But, the outcome is the same…a fixed-price hedge.

Table 7–1. Simple, fixed-price oil hedge using NYMEX futures contracts.

1) Refiner buys (30) June NYMEX futures contracts on 02/15/19 @ $60.50
2) June contract expires on 05/21/19 - Refiner closes-out position (sells back)
3) Refiner buys physical crude oil at "market" (ARGUS posting) for June

Scenario 1
June NYMEX Crude Oil Contract settles @ $60.75
1) Refiner has to sell back NYMEX contracts @ $60.75 for a gain of +$0.25 per Bbl.
2) Cash market (ARGUS posting) is now $60.75 also. Physical loss of ($0.25 per Bbl.

Refiner							
Buy June NYMEX		Sell June NYMEX Settlement (1)		Financial Gain/Loss	Buy June Physical Market	June Physical Gain/Loss	Net Gain/Loss
$60.50		$60.75		$0.25	$60.75	($0.25)	$0.00

Scenario 2
June NYMEX Crude Oil Contract settles @ $60.25
1) Refiner has to sell back NYMEX contracts @ $60.25 for a loss of ($0.25) per Bbl.
2) Cash market (ARGUS posting) is now $60.25 also. Physical gain of +$0.25 per Bbl.

Refiner							
Buy		Sell June			Buy June	June	
June		NYMEX		Financial	Physical	Physical	Net
NYMEX		Settlement (1)		Gain/Loss	Market	Gain/Loss	Gain/Loss
$60.50		$60.75		$0.25	$60.75	($0.25)	$0.00
(1) Closing-out position on Last Day of NYMEX trading							

Table 7-2. Simple, fixed-price oil hedge using Swaps

1) Refiner buys (30) June crude oil Swap contracts on 02/15/19 @ $60.50
2) June NYMEX contract expires on 05/21/19 - Refiner closes-out position (sells back the Swap)
3) Refiner buys physical crude oil at "market" (ARGUS posting) for June

Scenario 1
June NYMEX Crude Oil Contract settles @ $60.75
1) Refiner has to sell back Swap contracts @ $60.75 for a gain of +$0.25 per Bbl.
2) Cash market (ARGUS posting) is now $60.75 also. Physical loss of ($0.25) per Bbl.

Refiner							
Buy		Sell June			Buy June	June	
June		NYMEX		Financial	Physical	Physical	Net
Swap		Settlement (1)		Gain/Loss	Market	Gain/Loss	Gain/Loss
$60.50		$60.75		$0.25	$60.75	($0.25)	$0.00
(1) Closing-out position on Last Day of NYMEX trading							

Scenario 2
June NYMEX Crude Oil Contract settles @ $60.25
1) Refiner has to sell back NYMEX contracts @ $60.25 for a loss of ($0.25) per Bbl.
2) Cash market (ARGUS posting) is now $60.25 also. Physical gain of +$0.25 per Bbl.

Refiner							
Buy June Swap		Sell June NYMEX Settlement (1)		Financial Gain/Loss	Buy June Physical Market	June Physical Gain/Loss	Net Gain/Loss
$60.50		$60.75		$0.25	$60.75	($0.25)	$0.00
(1) Closing-out position on Last Day of NYMEX trading							

The advantage of using swaps for hedging is that you can achieve the same price protection without actually having to buy or sell NYMEX contracts. The refiner in the NYMEX example would have had to pay for the June WTI futures contracts almost immediately. Additionally, you can work with brokers either by phone ("voice" brokers) or through an electronic trading platform such as the ICE.[6] Swaps also can be customized to fit volumetric needs that vary from the standard contract specifications of NYMEX.

Basis Swaps

Basis is "the difference between the spot price of an asset and its futures price."[7] As we have stated, futures contracts have set delivery locations. Yet these same commodities change hands at hundreds of locations throughout North America. In addition, products like crude oil vary greatly in their quality and composition. The specific gravity, sulfur content, and Reid vapor pressure (RVP)[8] can all be different depending on the actual source of the crude oil.

These variations from the standard financial contract, in quality and/or location in physical commodities, are known as *basis* and represent a form of financial risk exposure for counterparties when these differences do exist. If delivery is to take place at a point other than the designated futures contract location, transportation costs may have to be considered. Or where the quality is different from the contract specifications, adjustments to the price may be warranted. From a price perspective:

Basis = Spot (cash/physical) price − NYMEX (futures)

Crude oil markets handle this by providing "deducts" and "adders" for the variances from WTI in their bilateral sales contracts. These can take many forms and will spell out what price adjustments will be made for API[9] gravity, RVP, and sulfur, among other possibilities. There may also be a transportation component, such as "WTI less transport," which would represent the actual cost to deliver the oil to Cushing, OK, the NYMEX futures contract delivery point.

In the natural gas market, since there is a uniform pipeline quality standard, the only real risk posed is the delivery location vs. Henry Hub, the so-called basis differential. Once known as the "geographical location basis differential," it is now more commonly known simply as the *basis*. A financial instrument called a *basis swap* can mitigate this risk. While it operates in a similar fashion as a regular swap, its settlement is very different. This is another form of a *fixed-for-floating* swap, where the fixed portion is the current market value and the floating portion has to be determined at settlement.

As we saw above, swaps settle by comparing the initial NYMEX price at the time of execution to the final settlement price for that NYMEX futures contract. For natural gas, that sets the price equivalent at Henry Hub in southern Louisiana. However, for other delivery points, other financial instruments will be needed in order to hedge the risk.

Take for example a producer with natural gas in Oklahoma who wishes to hedge the price in a future month. A NYMEX swap can get the producer a fixed price at Henry Hub, but that does not represent the price for the producer in Oklahoma. So the producer must look to the basis swaps market for the actual differential in price between the two points. Otherwise, as the price differences between Henry Hub and Oklahoma vary, the producer is exposed to what is known as *basis risk*.

The value of a basis swap is determined based on several factors but is closely tied to the market's perception of the value of gas in each location for which a basis market exists. (These are usually most, but not all, of the physical pricing points found in the industry publications.) One would think that this differential would simply be the transportation cost to get the gas from Oklahoma to the Henry Hub. While this makes perfect sense, there are two main factors that prevent this from being the case.

- *Capacity*. There would have to be enough pipeline capacity to get *all* of the natural gas from Oklahoma to southern Louisiana for the transportation cost to be the only variable, which is not the case.
- *"Domestic" demand*. Oklahoma produces more gas than it consumes and is thus a net exporter of natural gas. However, at high periods of demand, there is less natural gas available to be sent out of state. Out-of-state markets will have to pay the prevailing prices if they wish to purchase natural gas from Oklahoma. Competition between these groups raises the price in Oklahoma. This tends to make the value of natural gas closer to that of Henry Hub, eliminating the idea of a purely transportation-based differential.[10]

In general, natural gas prices are higher in consuming areas and lower in the producing regions, since that is the normal direction of the flow of gas. As a result, prices are less in producing areas than in market areas. That means there is *normally* a price discount from the Henry Hub price for production zones and some type of adder to that same price for consuming zones. Since basis swap values reflect these price differences, they are generally negative for cash market producing locations and positive for cash market consuming locations.

The financial markets provide basis swap values, so commercial parties wishing to hedge their locational risk can readily find quotes. Those who provide basis swaps markets run complex models to establish the market values. NYMEX's *ClearPort* electronic platform and *ICE* each trade basis swaps, as do the traditional voice brokers at financial institutions.

So our Oklahoma producer, wishing to set a fixed price, will utilize both a NYMEX swap to establish a Henry Hub price and a basis swap to hedge the price differential for Oklahoma. As with any hedge, the commercial entity will always take a position in the financial market that is the opposite of its physical position. In this case, the producer would sell both the NYMEX swap and the basis swap.

Since basis swaps are tied to the cash market, the settlement price is tied to the indexes for the respective locations as published each month. The majority of basis swaps for natural gas settle against the location's index as reported in *Platts Gas Market Report*, a natural gas market publication described in the chapter on cash markets.

For our example, we will use a pricing point known as "Panhandle Eastern Pipe Line Co. for Texas, Oklahoma (mainline)." This is an interstate natural gas pipeline that runs through Oklahoma on its way to market areas in Kansas, Missouri, and the upper Midwest region. Because a large volume is transported on Panhandle Eastern Pipe Line Co. (PEPL), it functions as a "liquid" point for trading basis swaps in Oklahoma.

At the time the hedge is executed, the fixed-price swap is priced at the current NYMEX price and the basis swap is priced at the current value in the marketplace for the desired location. The producer would sell both the NYMEX swap and the PEPL basis swap to a financial counterparty. Using our producer example, let's assume NYMEX was at $3.50/MMBtu and the PEPL basis swap, based on the current market quotes, was Henry Hub less $0.25/MMBtu. The price for the producer is now set at $3.25/MMBtu as follows:

NYMEX + Basis = Locational fixed price

When the *Platts Gas Market Report* is published for that month, the PEPL index price is compared with the NYMEX final settlement for that same month. (The index price is subtracted from the NYMEX settlement price.) The difference is the "actual" basis swap value for that month and represents what was the floating part of the swap and is now the settlement for the basis swap. That is then compared to the basis swap value set previously, the $0.25 fixed portion of the swap. The producer settles with the counterparty in the same fashion as with a normal swap agreement. Table 7-2 illustrates how the settlement of a basis swap works.

Table 7–3. NYMEX Basis Swap for Natural Gas

Basis Example 1

1/15/2019	«Fixed»		Platt's	NYMEX	Actual
		PEPL Basis	February	Settlement	Basis
	Feb-19	($0.200)	$2.530	$2.750	($0.220)
		Sells	Buys	Pays	
Seller (Producer)	Feb-19	($0.200)	($0.220)	$0.020	
		Buys	Sells	Receives	
Buyer (Broker)	Feb-19	($0.200)	($0.220)	($0.020)	

Basis Example 2

1/15/2019	«Fixed»		Platt's	NYMEX	Actual
		PEPL Basis	February	Settlement	Basis
	Feb-19	($0.200)	$2.530	$2.780	($0.250)
		Sells	Buys	Receives	
Seller (Producer)	Feb-19	($0.200)	($0.250)	$0.050	
		Buys	Sells	Pays	
Buyer (Broker)	Feb-19	($0.200)	($0.250)	($0.050)	

In order to fully comprehend how the settlement process for basis swaps works, we need to cover the addition and subtraction of negative numbers. For the purposes of simplicity, picture a number line where Henry Hub is the zero point. Generally, as you move along to locations north and east of Henry (to the right on the number line), you are in consuming areas where the value of gas increases, and thus prices are generally Henry "plus." That would make the basis swap values more "positive" than Henry Hub along the number line. Conversely, supply areas generally have lower prices than Henry, and their basis swap values are more "negative." (The exceptions here are prices for the large consuming regions in the Southwest and West. As gas moves west out of Texas, for example, demand exceeds supply and prices start to increase, which increases the basis swap values.)

Areas on the positive side of the number line traditionally have included regions such as the Northeast,[11] Southeast, upper Midwest, Southwest, and West. On the negative side are mainly the supply regions of the Mid-continent, Rocky Mountains, West Texas/New Mexico, and the Pacific Northwest.

As basis swap values become more negative relative to Henry, we refer to this as *widening*, i.e., the distance away from the zero point is increasing. If the opposite is true and values are getting closer to Henry, we refer to this as *tightening*, i.e., the distance away from the zero point is decreasing. Keep in mind that these basis swap fluctuations represent an increase or decrease in the fixed price for each location since they are added to the NYMEX price to establish the locational price. Widening basis swap values mean higher prices on the positive side of the number line but lower prices on the negative end. Conversely, tightening basis swap values represent lower prices on the positive side of the number line but higher prices on the negative end (the negative values are decreasing, leading to prices that are closer to those at the Henry Hub). Table 7–4 illustrates this concept using basis swap values and a NYMEX Henry Hub price to establish the locational fixed prices.

Table 7–3. Basis swap values at Henry Hub

Basis Swap	($0.25)	($0.20)	($0.15)	($0.10)	$0.00	$0.10	$0.15	$0.20	$0.25
Fixed-price	$3.25	$3.30	$3.35	$3.40	$3.50	$3.60	$3.75	$3.70	$3.75

Now, going back to our example above, the producer first *sold* the basis swap at –$0.25. When the actual basis value was calculated, it was a –$0.20, which is a greater value on the number line as it represents a lower discount to Henry. In this case, then, the producer *bought* the basis swap back at higher value and thus lost money: –$0.25 – (–$0.20) = –$0.05. On the other side of this transaction, the broker *bought* the basis swap from the producer at –$0.25 and *sold* it back at –$0.20, making $0.05: –$0.20 – (–$0.25) = $0.05.

Figure 7–1 shows actual North American natural gas basis swap quotes from CME's ClearPort electronic over-the-counter platform.

Description ↓	Prev Close
Feb 19 NYMEX CIG Rockies Natural Gas (Platts IFERC) Basis Futures Clearport	-0.2200
Feb 19 NYMEX Columbia Gas Transmission Corp. Appalachia (TCO) Basis (Platts IFERC) Futures Clearport	-0.2240
Feb 19 NYMEX Demarc Natural Gas (Platts IFERC) Basis Futures Clearport	-0.0900
Feb 19 NYMEX Dominion South Point Natural Gas (Platts IFERC) Basis Futures Clearport	-0.2440
Feb 19 NYMEX Houston Ship Channel Natural Gas (Platts IFERC) Basis Futures Clearport	0.0570
Feb 19 NYMEX MichCon Natural Gas (Platts IFERC) Basis Futures Clearport	-0.0880
Feb 19 NYMEX Panhandle Natural Gas (Platts IFERC) Basis Futures Clearport	-0.2510
Feb 19 NYMEX Permian Natural Gas (Platts IFERC) Basis Futures Clearport	-1.1850
Feb 19 NYMEX Sumas Natural Gas (Platts IFERC) Basis Futures Clearport	0.4040
Feb 19 NYMEX Transco Zone 6 Natural Gas (Platts IFERC) Basis Futures Clearport	5.8220
Feb 19 NYMEX Waha Natural Gas (Platts IFERC) Basis Futures Clearport	-1.4730

Fig. 7–1. Natural gas Basis Swaps quotes, NYMEX *ClearPort*

Note the description. The month is shown followed by the exchange (NYMEX). Then the physical pricing point is listed, along with the published settlement index reference. (ClearPort is still using the traditional name associated with *Platts Inside FERC* index publication, now known as *Platts Monthly Price Guide*.) The financial instrument, "Basis Futures," and the "sub-exchange," ClearPort, is shown. The "Previous Close" column represents the latest trading day's settlement value for the basis swap at the given location.

Several key basis swaps locations were chosen to give the reader an idea of the variances in differentials from the NYMEX natural gas contract delivery point at the Henry Hub in southern Louisiana. Some represent mainly supply areas, while others are large consuming regions.

- *CIG Rockies*. This point represents the physical prices for natural gas that originates in the Rocky Mountain region and is transported via Colorado Interstate Gas Company. Gas from this region can move East, West, SE and, NW. The area is a net exporter of natural gas.
- *Columbia Gas Transmission*. This is a mainline transmission pipeline that serves portions of the Northeast and runs through the Marcellus Shale.
- *Demarc*. This is a "line of demarcation" on Northern Natural Gas Company's main transmission line running through Kansas. It delineates the Supply area from the Market area in the rate matrix.
- *Dominion South*. This represents a point on Dominion Gas Transmission's mainline system in SW Pennsylvania/NW West Virginia. It is reflective of gas prices in the Marcellus Shale and is, at times, constrained due to limited takeaway capacity in the region.
- *Houston Ship Channel*. Gas delivered here serves the huge crude and petrochemical refining corridor along the Gulf Coast in eastern TX.
- *MichCon*. MichCon is the abbreviation for Michigan Consolidated Gas, a large Midwest utility. This pricing point reflects market demand in the upper Midwest. The upper Midwest has access to gas supplies coming

from the Gulf Coast, Mid-continent, Rocky Mountains, and from both the Marcellus and Utica Shale basins, thus providing substantial gas-on-gas competition for market share.
- *Panhandle.* Short for Panhandle Eastern Pipeline Company, this interstate pipeline delivers gas to the Midwest and upper Midwest from sources in the Texas Panhandle, Oklahoma, and Kansas. A large enough volume moves through the region on Panhandle that the financial markets found solid liquidity for the basis swaps market using this pipeline. The value reflects the supply area in TX/OK/KS. It is negative since the region is a net exporter of natural gas, even in winter.
- *Sumas.* This "hub" represents the value of natural gas entering the United States from Canada into Northwest Pipeline at the Washington State border. (At +0.40, the market clearly expects cold in northwestern United States.)
- *Permian.* This is an area in West Texas which encompasses the Permian Basin producing area. As of this writing, more natural gas is being produced than pipeline capacity can handle. Thus, the large discount to Henry Hub seen.
- *Waha.* Several pipelines crisscross West Texas near the Midland area. This creates a "hub" for the transfer of gas from one pipeline to another. Prices here reflect the supply/demand dynamics for intrastate TX (west to east), as well as for gas going west into NM, AZ, and CA. A large discount to Henry Hub applies to this area as well for the same reasons as in the Permian Basin.
- *Transco Zone 6.* This delivery area represents a large market area in the Northeast as well as the Marcellus Shale supply basin. As you can see, the price here is also higher than Henry Hub, which would be normal for a winter month. However, vast amounts of the Marcellus Shale gas still cannot get the East Coast to supplement the demand in this area, or the basis value would be less. Again, pipeline projects and "backhaul" arrangements are in the works that should shift this dynamic.

There are also Swaps for crude oil with the most widely-quoted being those that represent differentials between the WTI Cushing, OK, delivery point and, the Permian Basin and, Houston Ship Channel areas.

Using Basis Swaps to Hedge Transportation

Basis swaps can also be used to hedge the cost of natural gas transportation via pipelines since, in essence, we are looking at the value of one location (receipt) vs. another (delivery). We have addressed the fact that basis swaps are the difference between the price of natural gas at the Henry Hub and other locations. So each location's value when compared to another has a similar "constant" relationship that being, the Henry Hub price. As a result, the difference in basis swaps from one location to another represents the market value placed on the transportation between those two points, or the *spread*.

If we need to determine the fixed price of gas entering a pipeline on the supply end, we know we need the NYMEX Henry Hub price and the applicable basis swap value. We will also need the NYMEX Henry Hub price and basis swap value at the delivery point to establish the fixed price there. So we would have something that looks like this:

Transportation spread = NYMEX + Basis (market price) − NYMEX + Basis (supply price)

If we cancel out the NYMEX, we are left with the market basis *less* the supply basis.

For example, using the basis swaps quotes shown in figure 7–1, we see that gas in Oklahoma (Panhandle) is valued at Henry Hub less ($0.25), while gas delivered to the upper-Midwest (MichCon) is valued at Henry Hub less ($0.09). If we assume a NYMEX price of $3.50/MMBtu, we have an equation that looks like this:

$3.50 + ($0.09) − ($3.50 + ($0.25)) = $3.41 − $3.25 = $0.16

If we remove the constant, $3.50/MMBtu, we are left with a total of +$0.16 of value difference between the two geographic points:

($0.09) − (−$0.25) = $0.16.

Should the actual cost of transporting gas from Oklahoma to Michigan be ≤ $0.16, a shipper can sell the Panhandle basis swap and buy the MichConl basis swap, thereby locking in this differential without having to also execute futures, forwards, or swaps to establish fixed prices as well.

Summary Points

1. "Swaps" are another form of financial derivative which can be used for trading or hedging.
2. Swaps are an exchange of payments between two parties.
3. They are strictly financial. No physical commodity deliver takes place.
4. One party to a Swap transaction agrees to pay a current market price ("fixed") while the other agrees to pay a price in the future ("floating"/settlement).
5. Basis Swaps represent the difference between a delivery point for a financial derivative contract and a different physical transaction point.

Review Exercises

1. _____ are exchanges of payments between two parties. They are strictly financial. No physical exchange of the commodity takes place.
2. One party to the transaction agrees to pay the current _____ price, while the other agrees to pay a _____ price in the future.
3. Swaps are non-_____ traded instruments.
4. They are a simpler and less expensive way to hedge _____ risk.
5. They can be used for hedging or _____.
6. Fixed-for-floating swaps use current _____ market prices and final settlement prices to determine the balance of payments under the agreements.
7. One very important swap is a "_____" swap, which is a market-determined value that represents the difference between the NYMEX Henry Hub and other natural gas trading points in North America.
8. Swaps trade in the _____ market, which can be an electronic platform like NYMEX's ClearPort, the Intercontinental Exchange (ICE), or via voice broker.
9. Basis swaps can be used to hedge the price spread exposure inherent in a _____ agreement.
10. Variations from the standard financial contract, in quality and/or location in physical commodities, are known as _____ and represent a form of financial risk exposure for counterparties when these differences do exist.

Notes

1. Steven Errera and Stewart L. Brown, *Fundamentals of Trading Energy Futures and Options*, 2nd ed. (Tulsa: PennWell, 2002).
2. Hull, John C., *Options, Futures, and other Derivatives*, 9th ed. (London: Pearson Education)
3. Iris M. Mack, *Energy Trading and Risk Management: A Practical Approach to Hedging, Trading and Portfolio Diversification* (Singapore: John Wiley & Sons, 2014).
4. Ibid.
5. Ibid.
6. NYMEX ClearPort has a Henry Hub natural gas "penultimate" swap, representing the daily settlement one day prior to final settlement.
7. Hull, John C., *Options, Futures, and other Derivatives*, 9th ed. (London: Pearson Education)
8. *Reid vapor pressure* is the propensity for a liquid hydrocarbon to vaporize. In the case of crude, it is important for determining how to transport it, as well as the actual volume delivered as some product may evaporate.
9. The American Petroleum Institute establishes industry-adopted standards for crude oil and refined products. See www.api.org.
10. This tug-of-war between domestic Oklahoma end users and those out of state was never more apparent than during the winter of 2013/2014, when a daily price on ICE traded $50/MMBtu for gas in the state.
11. With the abundant gas supplies coming out of the Marcellus and Utica Shales, gas in the northeastern United States is actually seeking markets to the west, south, and southeast. This is making the basis values there negative to Henry Hub in some cases.

8

Financial Energy Derivatives: Spreads

Key Learning Points: Spreads

- Spreads are merely price differences between commodities that are interrelated somehow, have differing locations, or representing different months of the same commodity.
- Spreads are traded for hedge purposes (to reduce price risk) or purely for profit (speculation on price spread movement).
- The price differences are important, not the absolute prices themselves.
- Spread categories are *intramarket* and *intermarket*.
- *Intramarket* spreads involve price differences between locations or time periods for the *same* commodity.
- *Intermarket* spreads involve price differences between different, but related, commodities.
- Spread types include **time**, **location** and, **commodity**.
- Common spreads used for hedging:
    - "Crack" = crude oil vs. gasoline and heating oil
    - "Frac" = natural gas vs. NGLs
    - "Spark" = natural gas vs. electricity
    - Locational = WTI vs. Brent

Spreads

Commercial participants in the energy financial derivatives markets are concerned about the price for their products or the cost of the energy needed to produce their products, or both. They represent the inputs and outputs of the energy marketplace.

Spread trading "is a technique that takes advantage of the relative price movement between futures contracts."[1] Spread trading can be used for hedging purposes or purely for trading. Speculators, looking for arbitrage[2] opportunities, use spreads strictly for profit. Commercial entities may also use the price relationship between and among commodities for risk reduction.

Those who use spreads for pure speculative purposes seek to take advantage of the price differences between or among commodities and are not interested in the commodities themselves. Furthermore, they are looking to make a profit on the movement of the spread and not the absolute prices themselves. If the two prices are moving closer to one another, this is known as a tightening or narrowing spread, as explained previously. If they are moving away from each other, the spread is widening.

Certain rules of thumb exist for the trading of spreads:[3]

Spread Rule 1: If the spread between two contracts narrows, a profit will occur if the lower-priced contract has been purchased and the higher-priced contract has been sold. A loss occurs when the spread narrows if the lower-priced contract was sold and the higher-priced contract purchased.

Spread Rule 2: If the spread between two contracts widens, a profit will occur if the lower-priced contract has been sold and the higher-priced contract purchased. A loss occurs when the spread widens if the lower-priced contract is purchased and the higher-priced contract sold.

To put it in more simple terms, traders will execute the spreads based upon their perception of where prices may go. Errera and Brown state the following:[4]

1. If spreads are expected to narrow, buy low and sell high.
2. If spreads are expected to widen, buy high and sell low.

Let's look at an example of executing natural gas spreads with the following market data:
- NYMEX December 2018 NG price of $3.25/MMBtu
- NYMEX January 2019 NG price of $3.35/MMBtu

If one thinks this spread will narrow:
- Buy the $3.25/MMBtu contract.
- Sell the $3.35/MMBtu contract.

If one thinks this spread will widen:
- Sell the $3.25/MMBtu contract.
- Buy the $3.35/MMBtu contract.

Again, it is the *spread* between the prices that matters, and not the absolute prices themselves.

In spread trading, futures, forwards, or swaps can be used to achieve the desired results. A buy/sell is offset by a corresponding sell/buy. There are two main types of spreads: intermarket and intramarket.

Intermarket spreads are the "simultaneous purchase and sale of different, but related, commodities that have a reasonably stable relationship to each other."[5] They are also known as *intercommodity spreads*, which consist of a long position in one commodity and a short position in a different, but related, commodity.

Examples of different spreads are as follows:
- *Crack spread.* Buy crude oil and sell heating oil and/or gasoline (HO are RB are "cracked" from CL). This is also known as a *refiner's hedge*.
- *Frac spread.* Buy natural gas/sell propane. Midstream natural gas companies process natural gas into propane and other NGLs, so they buy the natural gas and sell the propane. This is not to be confused with *fracking*, or the fracturing techniques used during drilling. In this case, "frac" is short for "fractionation," the process whereby a composite stream of NGLs is separated into its different components or "purity" products. *Frac spreads* represent the difference between the cost of natural gas needed to produce natural gas liquids (NGLs) and the revenue from the sale of the NGLs, on a per-Btu basis.
- *Spark spread.* Buy natural gas/sell electricity. Electric generators can use natural gas to produce power. Spark spreads represent the difference between the cost of natural gas used to produce electricity at a power plant and the revenue derived from the sale of the power.
- *Heating oil vs. gas oil.*
- *Intramarket spreads.* Also called *intracommodity spreads*, these spreads are "the simultaneous purchase/sale of futures contracts on the same commodity for different delivery months."[6]

The following are several examples of different types of spreads:
- *Time spread* (also called *storage* or *carrying charge*)
  - Buy a natural gas contract in May/sell it in January.
  - Buy a heating oil contract in April/sell it in December.
- *NYMEX vs. ICE.* Applies for the same commodities (e.g., futures vs. swaps).
- *Locational.* Henry Hub vs. Panhandle natural gas (basis spread), or WTI vs. Brent crude.

In addition to traders who are merely interested in price movement to make money, commercial entities can use spreads to hedge their price risk. For example, as mentioned above, a crude oil refiner can buy crude contracts (hedge the price of feedstock) and sell heating oil and unleaded gasoline contracts (refined output) to establish a gross profit margin or *crack spread*.[7]

The most common crack spread is the *3-2-1 spread*, which represents the theoretical spread between 3 bbl of crude and its refined products, 2 bbl of gasoline and 1 bbl of heating oil.

Here is how that would be calculated where 1 bbl of gasoline (RBOB) = 42 gal and 1 bbl of heating oil (HO) = 42 gal.

Assumptions: CL = $55/bbl; RBOB = $1.60/gal; and HO = $1.95/gal.

Steps:
1. 3 bbl of crude @ $55/bbl = $165 (cost of crude).
2. 2 bbl of RBOB = 84 gal @ $1.60/gal = approximately $134 (revenue from RB sale)
3. 1 bbl of HO = 42 gal @ $1.95/gal = approximately $82 (revenue from HO sale)
4. Total refined product revenue ($134 + $82) = $216; less the cost of CL ($165) = $51. $51/3 bbl = $17/bbl spread.

The Hedging of Storage Capacity Using Time Spreads

A commercial entity with storage capacity can also use an intramarket (time) spread to hedge its seasonal risk. Heating oil or natural gas can be bought in the off-season and stored for resale during the peak winter months. (Most state public utility commissions require utilities to have emergency supplies on hand by the start of the winter season.)

Physical marketing and trading companies will often utilize storage capacity for arbitrage opportunities. Their "risk" amounts to the total cost to store the commodity vs. the spread between the purchase and sale months. This, too, can be hedged by using financial derivatives. The example below assumes a simple, seasonal storage contract, whereby the lessee can only inject gas into storage in a stated month and withdraw it in a subsequent (also stated) month. (Hedging using options for more flexibility is discussed in chapter 9.)

At this point, we do need to talk about two market price conditions that can exist, "contango" and "backwardation". A market is in *contango* if each successive month is priced higher than the preceding one. It is also known as a *premium* or *carrying charge market* since the costs of storing and, the opportunity costs, must be carried until the inventory is sold. A contango market is a storage market since the futures prices closer at hand are lower than those further out, allowing for a profit to be made on the seasonal arbitrage. Conversely, a market is said to be *backwardated* if the prompt month contract is higher than the successive ones. This condition normally occurs when there is some fundamental event driving higher demand in the near-term period.

A natural gas storage transaction can serve as an example. In this instance, April Henry Hub natural gas = $3.15/MMBtu, and January Henry Hub natural gas = $3.65/MMBtu. The following steps would be taken:

1. Buy April natural gas contracts @ $3.15/MMBtu.
2. Sell January natural gas contracts @ $3.65/MMBtu.
3. Gross "spread" = $0.50/MMBtu.
4. Deduct monthly storage fees (as set by storage lessor) and cost of capital.*

   *Based upon the standard terms for a natural gas bilateral purchase and sale contract,[8] the natural gas purchased in April will have to be paid for by the end of May. And revenue for the January sale will not be received until the end of February. There is an opportunity cost for the initial cash outlay for the supply, and most companies attach an internal interest rate cross-charge to that.
5. Net revenue × Volume contracted = Transaction revenue.

There are advantages to using spreads:

1. The risk exposure is limited to the change in the spread and not the absolute prices themselves. This requires less margin/credit assurance, as they are less risky than outright futures.
2. Commercial entities can hedge their risk using a simple financial derivative as opposed to entering into multiple contract positions.

Summary Points

1. Spreads are merely price differences between commodities that are interrelated somehow, have differing locations, or represent different months of the same commodity.
2. Spreads are traded for hedge purposes (reduce price risk) or outright trading (speculate on price spread movement).
3. The price differences are important, not the absolute prices themselves.
4. Spread categories are intra-market and inter-market.
5. Spread types are **time**, **location** and, **commodity**.
6. Intra-market Spreads involve the price differences between locations or time periods for the same commodity.
7. Inter-market Spreads involve the price differences between different but, related, commodities.

Review Exercises

1. _____ are merely price differences between commodities that are inter-related somehow, have differing locations, or represent different months of the same commodity.
2. They are traded together for _____ purposes (to reduce price risk) or outright _____ (speculation on price spread movement).
3. There are two main categories: _____ and _____.
4. Spread risk is lower than outright _____ positions.
5. The spread is the key, not the absolute _____.
6. The most common types of spreads are _____, _____, and _____.
7. _____ spreads can be used by oil refiners, midstream natural gas companies, and electricity generators to lock in or hedge gross margin.
8. Give an example of an intra-market spread: _____ vs. _____.
9. Give an example of an inter-commodity spread: _____ vs. _____.
10. Define the following spread types:
    a. "Crack" = _____ vs. _____ and _____
    b. "Frac" = _____ vs. _____
    c. "Spark" = _____ vs. _____

Notes

1. Steven Errera and Stewart L. Brown, *Fundamentals of Trading Energy Futures and Options*, 2nd ed. (Tulsa: PennWell, 2002).
2. Ibid. According to Errera and Brown, arbitrage is "the simultaneous purchase and sale of similar or identical commodities in two different markets in hopes of gaining a profit from price differentials."
3. Ibid.
4. Ibid.
5. Ibid.
6. Ibid.
7. Oil refineries "crack" the very large crude oil hydrocarbon molecules into refined products using heat, steam, or chemical catalysts.
8. The North American Energy Standards Board (NAESB) has a standard form bilateral contract for natural gas. See www.naesb.org.

9

Financial Energy Derivatives: Options

Options are another financial derivative used for speculative trading or hedging. They are called *options* because, in all cases, the buyer will have the "option" of exercising the derivative. They are referred to as *asymmetrical* since *the buyer has the right, but not the obligation*, to exercise the option. According to Mack, "An energy option is a derivatives instrument whose underlying asset is some type of energy product such as oil, natural gas, or electricity. It represents a contract sold by one party (options writer) to another party (options holder)."[1]

Key Learning Points: Options Contracts

- Options give the buyer the *right*, but not the *obligation*, to buy or sell financial energy contracts at some point in time in the future at a set volume and price. They are traded on both the NYMEX and over-the-counter markets.
- Options are much cheaper than outright contracts or swaps in that premiums usually represent only a fraction of the face value of the underlying contracts.
- As a result, a substantial volume of contracts can be "controlled" relatively cheaply.
- Options contract components list the commodity, volume, date, price (or *strike*), and premium to be paid.
- A *call option* gives the buyer the right to *buy* contracts at a fixed price, which creates a maximum price or *ceiling price*. These are mostly used by consumers of the energy commodity wishing to cap their price risk exposure.
- A *put option* gives the buyer the right to *sell* contracts at a fixed price, which creates a minimum price or *floor price*. These are mostly used by producers of the energy commodity wishing to limit their downside price risk.
- Options values (premiums) are calculated using algorithmic models.
- The most popular model is the Black-Scholes model.
  The components of an energy commodity options contract include the following:
    - Option type (call/put)
    - Commodity (underlying)

- Date (month/months; expiration date)
- Strike price (price at which the contracts can be bought or sold by buyer)
- Premium (cost)

Options

The options contract gives the holder the *right*, but not the *obligation*, to buy or sell the underlying energy contracts at the agreed-upon price (called the *strike price*) during a stated period of time or on a particular date (exercise date).

Car insurance is a good example of an option, or more specifically, a "Call Option". A *premium* is paid and the insured is covered for liability related to an accident, as well as damage, loss, or theft.

If an accident occurs, the insured (the buyer) has the right to "call" the insurance agent and exercise rights specified under the insurance contract. The "price" the insured will have to pay for the damages is limited to the amount of the deductible (strike price). If, for example, the term is one year and no claim is made, the "option" expires worthless (i.e., no payout is made by the insurance company since no claim was made). The insured's maximum exposure is the deductible, thereby establishing a "ceiling price." The premium is calculated using complicated mathematical models (actuarial tables, statistics, and probabilities).

This insurance scenario thus contains the main elements of an options contract:
1. Commodity—car insurance
2. Date—term of coverage
3. Strike price—deductible
4. Premium—cost of coverage

Energy options are very similar in nature. As with most financial derivatives, they can be used for hedging price risk or for outright trading. One key difference is that options represent the buyer's right, but not the obligation, to buy or sell futures/forwards contracts. (The options contracts themselves are not futures or forwards contracts but rather *the right to buy or sell those contracts*.) They are traded on regulated exchanges, as well as in over-the-counter markets. The buyer is under no obligation to purchase or sell the underlying commodity contracts if the pricing makes no sense.

The two basic types of options are Puts and Calls:
1. *Put options.* The buyer of the put option has the right, but not the obligation, to sell the underlying commodity's contracts at a set price, thus establishing a *floor price*.
2. *Call options.* The buyer of the call option has the right, but not the obligation, to buy the underlying commodity's contracts at a set price, thus establishing a *ceiling price*.

The elements of an energy options contract are the following:
1. *Options type.* The option will be either a Put or a Call. A put is the right to sell the underlying commodity; a call is the right to buy the underlying commodity.
2. *Strike price.* Also known as the *exercise price*, it is the stated price for which the underlying asset may be purchased (call option) or sold (put option) by the holder (buyer) of the option if the choice is made to exercise the option.
3. *Commodity.* The stated underlying asset that is to be bought or sold.
4. *Date.* The month or months for which the underlying commodity may be bought or sold up to and, including, the option's settlement date.
5. *Premium.* The price of the option, which is paid to the option seller (writer) by the option buyer. It represents the buyer's maximum exposure.

Option types are the following:
- *Calls.* These give the buyer the right, but not the obligation, to *buy* the underlying financial energy contracts should the market price exceed the strike price of the option contract. In that case, the buyer would call the seller of the option and request the contracts.
- *Puts.* These give the buyer the right, but not the obligation, to *sell* the underlying financial energy contracts should the market price fall below the strike price of the option contract. In that case, the buyer would put the contracts to the seller of the option, who must purchase them.

There are two primary settlement types:
- *American.* This settlement type can be exercised at any time before the expiration date.[2]
- *European.* This settlement type can only be exercised on the expiration date (most common).

The risk exposure for the buyer of an option is merely the cost of the option, i.e., the premium. The buyer will never pay more than that. On the other hand, the seller, or *writer*, of an option bears all the risk and is exposed to any price movement above the strike price of the call option and below the price of the put option.

Since only a premium is paid up front, one of the main advantages of options is that the buyer of the options can control a large number of contracts for a small price.[3]

For example, with a call option, the buyer is not buying the underlying contracts outright, but instead is buying the *right* to purchase them at a set price (strike price). Thus the buyer could have the right to buy 100 contracts and only have to pay the premium for the option and not the total cost of 100 outright futures contracts. (Premiums are priced per unit.)

Example of crude contract options:

1. To buy 100 crude contracts (100,000 bbl) @ $50.00/bbl = $5,000,000.
2. A call option for 100 $50.00 crude contracts (100,000 bbl) @ $2.00 premium/bbl = $200,000.
   a. The buyer now has *the right to purchase* 100 crude contracts @ $50.00/bbl but has only paid $200,000 at this time.
   b. If crude prices exceed $50.00/bbl, the buyer can call on the actual contracts. At that time, the buyer will have to pay the full price for the contracts ($5,000,000). So, the buyer essentially bought price insurance on 100,000 bbl for $200,000.

Note: If not exercised, options expire worthless. Furthermore, options are time-sensitive. The closer to the expiration date, the less value the option has. (There is less risk exposure with less time remaining.)

Options Models

There are numerous mathematical models that are used to determine options premium values. The most well-known is the Black-Sholes model. It is an extensive algorithm that only needs a few inputs to calculate an option's value. Inputs include the following:

- *Asset price* (current market price)
- *Strike price* (buyer's desired price)
- *Days to expiration* (of the underlying commodity contract)
- *Volatility*. Volatility of the underlying contract represents not only the magnitude of price changes but also the speed with which they change. This value can be obtained in the marketplace.
- *Interest rate*. This is the opportunity cost of paying the premiums up front vs. investing the cash in something else. The Federal Reserve's prime rate is normally used.

Once the information is entered into the model, call and put premium values are calculated. In addition, proprietary options models can be written by the sellers/writers of options, who generally employ quantitative analysts for that purpose.

Options models also produce theoretical values known as "the Greeks." Each of these measures the options' sensitivity to changes in asset value, volatility, and interest rates.[4]

- *Delta*. A measure of how much the value of an option should change when the price of the underlying commodity rises by $1.

- *Gamma.* Measures the rate of change in the delta for each one-point increase in the underlying asset.
- *Theta.* A measure of the dollar amount that an option will lose each day due the passage of time (also known as *time decay*).
- *Vega.* Measures the sensitivity of the option's price to changes in volatility.
- *Rho.* Measures how the option price changes with a change of one percentage point in the interest rate.

Perhaps the one Greek measure that receives the most attention is delta, since it indicates the value of the option relative to a $1.00 movement in the asset's price. For instance, a delta of 0.35 means that for every increase/decrease in the asset's price, the value of the option increases (call option) or decreases (put option) by $0.35. This can also then be used to calculate one's underlying contract position rights (buyer) or obligations (seller). So when determining one's overall open volumetric position, this figure is included, in addition to any actual contracts held/sold. At inception, the seller's delta is assumed to be 0.50.

Characteristics of Options

Options can be described as *in-the-money (ITM), at-the-money (ATM),* and *out-of-the-money (OTM),* all of which are determined by the relationship of the strike price to the current asset (market) price.

Intrinsic value is the way to measure ITM, ATM, or OTM status. It is the amount by which an option is in-the-money. The intrinsic value of a call option is the difference between the asset (underlying) price and the option's strike price. For example, if a call option's strike price is $60.50 and the current asset price (market) is $61.50, the call option has an intrinsic value of $1.00. Conversely, the intrinsic value of a put option is the difference between the strike price and the asset (underlying) price. If the strike price is $61.50 and the asset price (market) is $60.50, the intrinsic value of the put option is $1.00.

Any option is said to be in-the-money if it has intrinsic value, at-the-money if the strike prices equals the asset price, and out-of-the-money if it has no intrinsic value. Call options are OTM if the price of the underlying asset is less than the strike price. Put options are OTM if the price of the underlying asset is greater than the strike price.

Hedging Using Options

So how are options contracts used for hedging? Since this book is intended to be a first primer on energy derivatives, we are going to work with "plain vanilla" options. These are options that have standard terms and no unusual features or special conditions.

Let's take a crude oil refiner as an example. The company is concerned about rising crude oil prices. But rather than go out and buy hundreds of futures contracts and lock in the price now, the company decides to purchase a call option at a strike price that limits its exposure to rising prices. In doing so, the company establishes a maximum (ceiling) price and will never pay more than that.

Looking at May futures prices, the company decides to buy crude oil options as a hedge against the potential for higher prices moving into the summer driving season. When the company enters into the call option contract, the May futures price for crude is $50.00/bbl. The company wants to make sure its price does not exceed $55.00/bbl, so it buys a call option with a strike price of $55/bbl. If May prices remain below $55.00/bbl, the refiner does nothing and is out only the premium. However, should May prices exceed $55.00/bbl, the refiner calls the option seller and requests the number of crude oil contracts agreed upon at the $55.00/bbl strike price. In this scenario, the refiner will never pay more than $55.00/bbl for its crude supply. Furthermore, the refiner captures all the downside should prices fall.

On the flip side, let's consider a crude oil producer concerned about falling prices. This producer enters into a put option for May to establish a floor price and chooses a $45.00/bbl strike price, thus establishing the lowest price at which the producer will have to sell its crude oil. Should prices fall below that level, the producer will contact the options seller and request the right to sell the underlying financial contracts at $45.00/bbl. Should prices remain above $45.00/bbl, the producer would do nothing and is out only the premium. In this way, the producer can reap all the benefits of higher prices regardless of how high they go.

The producer can also offset the cost of the put option by selling a call option at a strike price where the premium is the same, or nearly the same, as the premium paid for the put. If the two premiums are the same, this transaction is known as a *costless collar* or a *no-cost collar* because the price the producer will receive is "collared" between the floor (put strike) and ceiling (call strike) prices. In this scenario, the producer is willing to give up any upside above the call strike price in order to have the downside protection (put strike) without paying a premium. This is an important consideration when a producer looks to utilize options, since the cash outlay for premiums may be better put to use as capital for E&P activities.

Table 9–1 provides some good examples of possible "zero-cost" collar scenarios. (Option strike prices are shown down the middle of the table, with put premium quotes shown on the left and call premium quotes on the right.) A producer could buy a $50.50/bbl put option for $1.06 and then sell a $54.50/bbl call option for about same price @ $1.08. The producer has made a business decision to gain downside protection if prices fall below $50.50/bbl. But, the producer is giving up any upside potential above $54.50/bbl in order to pay nothing for setting the floor price. Thus, the producer's price is "collared" between $50.50 & $54.50.

Table 9–1. NYMEX crude oil option quotes.

Last	Strike	Last
Mar 2019	52.61	last 52.61 volume 673928
0.42	47.50	5.53
0.49	48.00	5.11
0.58	48.50	4.69
0.68	49.00	4.29
0.79	49.50	3.90
0.92	50.00	3.53
1.06	50.50	3.18
1.23	51.00	2.85
1.41	51.50	2.53
1.62	52.00	2.24
1.84	52.50	1.96
2.09	53.00	1.71
2.36	53.50	1.48
2.65	54.00	1.28
2.96	54.50	1.08
3.29	55.00	0.92
3.64	55.50	0.76
4.01	56.00	0.64
4.40	56.50	0.53
4.81	57.00	0.44
5.23	57.50	0.36

Source: Marketview.com.

Summary Points

1. Options are another financial derivative used for speculative trading or hedging.
2. Options give the buyer the *right*, but not the *obligation*, to buy or sell financial energy contracts at some point in time in the future at a set volume and price.
3. A *call option* gives the buyer the right to *buy* futures contracts at a fixed price, which creates a maximum price or *ceiling price*.
4. A *put option* gives the buyer the right to *sell* futures contracts at a fixed price, which creates a minimum or *floor price*.
5. An energy producer, seeking to obtain downside price protection, without having to layout cash for a Put premium, can enter into a "costless collar". That entails the producer selling a Call option at a premium that offsets the cost of the premium to be paid for the Put.

Review Exercises

1. _____ are a simple and less costly way to hedge price risk than the outright purchase or sale of energy financial contracts.
2. They give the_____ the right, but not the obligation, to buy or sell the underlying energy commodity contracts at the strike price.
3. The main components of an options contract are:
   a. _____
   b. _____
   c. _____
   d. _____
4. The seller or _____ of the option assumes all risk.
5. Options can be used for hedging or outright _____.
6. Commercial entities concerned about rising energy prices, i.e., refiners, would enter into a call option, thereby establishing a maximum or _____ price for their commodity needs.
7. Commercial entities concerned about falling energy prices, i.e., producers, would enter into a put option, thereby establishing a minimum or _____ price for their commodity.
8. The _____ model is the most popular options valuation model.
9. The inputs necessary to run an options valuation model are:
   a. _____
   b. _____
   c. _____
   d. _____
   e. _____
10. If not exercised, options expire _____.

Notes

1. Iris M. Mack, *Energy Trading and Risk Management: A Practical Approach to Hedging, Trading and Portfolio Diversification* (Singapore: John Wiley & Sons, 2014).
2. The American-type settlement is the most popular exchange-traded option See John C. Hull, *Options, Futures, and Other Derivatives*, 9th ed. (Upper Saddle River, NJ: Pearson, 2015).
3. During the investigation into the causes of the 2007/2008 financial collapse in the United States, it was finally recognized just how much volume is represented by options, regardless of financial instrument. This "hidden" volume only served to exacerbate the problem. Dodd-Frank regulations attempt to shrink overall volumetric positions for financial derivatives, including the so-called options effect.
4. https://www.cmegroup.com/education/courses/option-greeks/options-the-greeks-options-premium-and-the-greeks.html

10

Technical Analysis

The price of a specific security at any one time is affected by the knowledge, hopes, fears, and expectations of all those people who already own it or who might be thinking of buying.

—Martin Pring

Overview

Thus far, we have addressed the fundamental factors that influence energy prices. And we established that there are two main groups that trade in the financial energy commodities markets, commercial and noncommercial. The latter group represents the "pure" traders or speculators. These participants are only interested in price movement. The type of commodity does not matter to them. In order to make trading decisions, they primarily use technical analysis as opposed to fundamental analysis.

Key Learning Points

- Differences exist between technical and fundamental market analysis, and it is important to be able to distinguish between them.
- Technical analysis relies on the principals of probability and statistics.
- It is important to be able to identify different types of technical charts and their uses.
- The three most popular technical charts are line (or *close only*), bar, and candlestick.
- Line charts record only the daily closing price and are best used for long-term trending.

- Bar charts indicate the daily high, low, opening, and closing prices for the trading day.
- Candlestick charts also show the high, low, opening, and closing prices but do so in a fashion as to illustrate the market direction, up or down, for that day's trading.
- It is important to recognize trend lines and market signals.
- Trend lines are used to indicate the trends.
- Identifying the current trend is the first step.
- Determining if the trend is going to change is the next step. This can be ascertained if the preponderance of the evidence indicates it will. Buy/sell decisions will then be made.
- Various price patterns exist for traders to identify.
- Volume is a good indicator of market activity and can reinforce the day's price movement.
- Resistance is the price level at which sellers enter the market again. It establishes a ceiling price in the current market.
- Support is the price level at which buyers enter the market again. It establishes a floor price in the current market.
- Moving averages are good studies to utilize if you believe in the statistical premise of *reversion to the mean*.
- It is important to be able to analyze "momentum" indicators.

Technical analysis involves the use of charts to track price movement, establish the current market trend, and determine the probability of prices moving in one direction or another. Simply put, technical or day traders are interested in market activity as illustrated by the resulting prices.

Since the prices that occur in the market are the result of human decision making, technical analysis really examines the behavior of market participants. According to Pring,

1. Prices in freely traded markets are determined by the attitudes of actual and potential market participants towards the emerging fundamentals.
2. These attitudes evolve in trends.[1]

As such, patterns emerge that have a high probability of recurring. It is precisely these events that technical traders are looking for. The idea is to "identify trend changes at an early point and ride a trend until the weight of the evidence points in the opposite direction."[2] But make no mistake: fundamental factors can cause traders to react emotionally, the results of which are also reflected in the price action.

In technical analysis, traders must first establish what the current price trend **is**: up or down. They must then determine the probability of the trend lasting or changing direction. It is this information that guides their buy/sell decisions.

There are several types of charting methods, but three of them are the most popular:

1. *Bar chart.* In this chart, a vertical line is shown for each time increment selected. In figure 10–1, a daily chart is used to show the December NYMEX contract for crude oil. Each bar shows the price results for that day's trading. The mark to the left of the bar represents the first trade of the day, or the *Open.* This is the price of the first trade that occurs right after the bell rings to start trading. The vertical line represents the full range of prices for the day, that is, the *High* and *Low* prices recorded. The mark to the right of the bar represents the final *Close,* or the settlement price for the day. This chart is often referred to as the *OHLC chart* (Open/High/Low/Close).

Fig. 10–1. Daily bar chart for NYMEX December crude oil contract

2. *Close-only.* This type of chart shows only the daily market settlement price. It provides much less information than the bar chart and is mainly used for longer-term trend analysis. Figure 10–2 shows the same December NYMEX crude oil contract in this form.

Fig. 10–2. Close-only chart for NYMEX December crude oil contract

3. *Candlestick.* These charts were developed by the Japanese centuries ago. They provide information similar to the bar chart but also indicate "up and down" days. That is, they clearly show the direction the market took on a daily basis. The top end of the "candle" still represents the high for the day, and the lower end represents the low, but the "body" indicates the open and closing prices in relation to one another. For example, if the open is higher than the close, the open price is at the top of the body of the candle and represents a day where prices fell (a solid body). Conversely, if the close is found on the top of the body, it represents an up day and appears on charts with a hollow body. By looking at the chart in figure 10–3, you can now see the up and down days are easily visible on the candlestick chart. By counting these, we can determine the current trend. For traders, the question is, When will it reverse course?

Fig. 10–3. A candlestick chart indicates the direction the market took on specific days.

Trend Lines

The longer a trend is in existence, the greater the implications of its reversal once a signal has been given.

—Martin Pring

As mentioned earlier, the first step in technical analysis is to identify the current trend. Once established, trends tend to perpetuate for an extended period, giving rise to the old Wall Street adage, "The trend is your friend." Trend lines can be used to identify both long- and short-term price trends. They are also used to indicate Support and Resistance prices and, "channels" (covered later). When drawing a trend line, it only has significance if it touches at least two price points. Figure 10–4 shows an obvious long-term downtrend going back two years to when crude prices began their dramatic decline. Figure 10–5 illustrates two short-term trend lines, one up and one down.

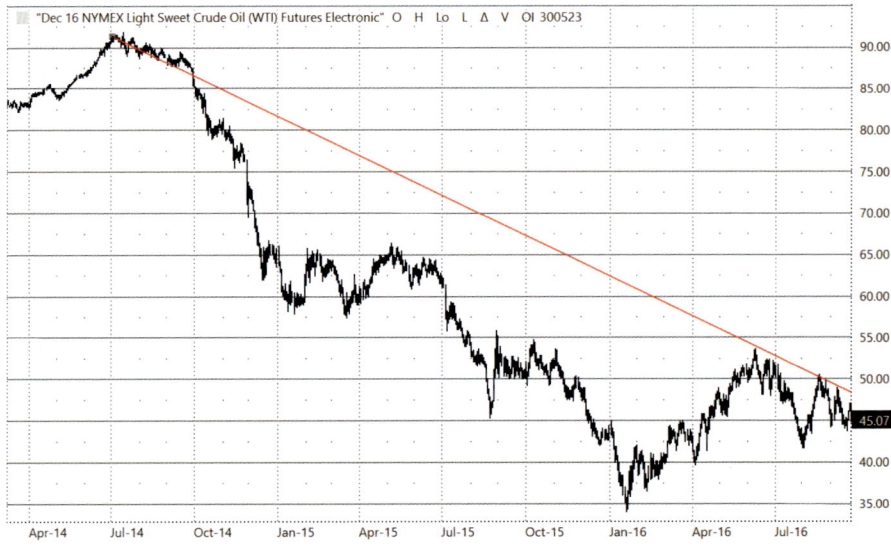

Fig. 10–4. Long-term trend line chart

Fig. 10–5. Short-term trend line chart

Pring suggests that there are four types of trends, designating them as short-term (3–6 weeks); intermediate (6 weeks to 9 months); primary (9 months to 2 years), and what he calls the *secular trend*. The secular trends, "comprising several primary trends, usually last between 15 and 20 years."[3] In energy commodities, liquidity beyond 5 years is very thin, and thus secular trends are not of much value.

Speculative traders will also seek to identify much shorter intraday trends as they look for opportunities to get in and out of positions in the same day. Obviously, these trends can change direction suddenly on the slightest bit of news or rumor. And one must keep in mind the prevailing longer-term trend, whether it is an entire day, a week, or a month. Bucking the trend does not make much sense and is akin to paddling up river.

Elements of a technical chart

Some common elements of a technical chart include the following:

1. *Volume.* One of the simplest clues to the strength of price movement is that of the volume of contracts traded. If a price shows a large range or change in direction on a particular day, looking at the volume of contracts traded indicates how well supported that move was by the market participants. A $0.10 movement up or down in natural gas is not very significant if a low volume of contracts has traded. On the other hand, large trading volumes definitely reinforce the price action for the day, almost as if those trading have agreed on the price outcome. Figure 10-6 is a Daily Bar chart with volume for natural gas. Notice that on January 14th, prices rose substantially and a very large number of contracts exchanged hands, solidifying the move. On January 22nd, prices fell, and high volume traded once again. Both of these volumes add legitimacy to the price action for those days.

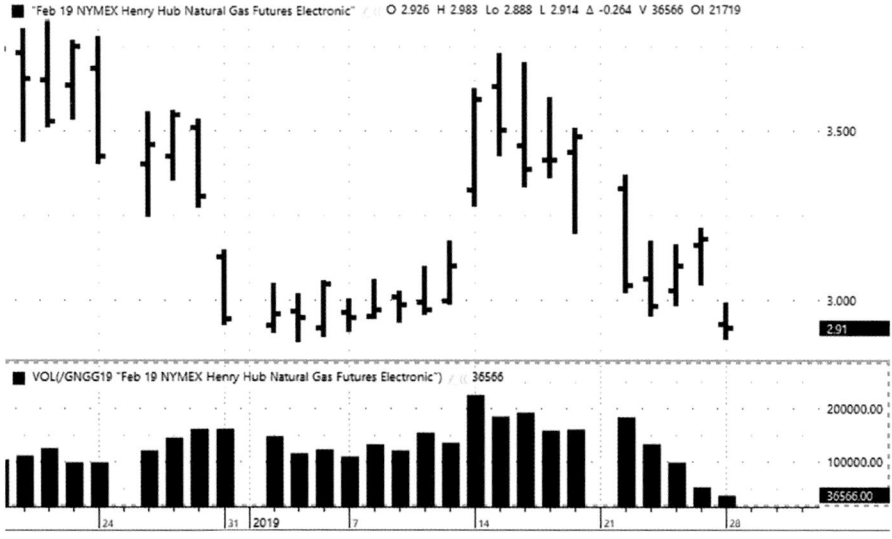

Fig. 10-6. Daily Bar Chart with Volume

2. *Moving averages.* Those who have studied statistics will be familiar with the term *reversion to the mean*. This concept hinges on the idea that all prices will eventually return to their average, despite dramatic movements up-or-down. I have found this to be especially true for energy commodities, at least in the short-term. Therefore, tracking commodity moving average prices can be a good signal for a change in the direction of a trend. ("Moving" averages simply means that, for a given period, the average price is calculated. Then, when the next trading day settles, that price is added to the time period and the first price of that period is dropped-off.) For instance, Figure 10-7, below, shows that the moving average (MA) for March 2019 crude oil over a 30-day period was $49.52 /bbl, while prices had risen to as high as $54/bbl. This means there is a good probability that they will eventually fall towards $49.52/bbl. It may be a gradual decline, which also means the average will change, but as long as the MA is lower, prices will gravitate toward it. The exact opposite occurs when prices fall below the MA. Figure 10-7 also illustrates this principal with the March 2019 crude oil contract. Note that the time frame for the MA is set to the particular trader's needs. I have set the MA at five days, as that represents a full week of trading (regular-session pit trading only occurs on weekdays). Note how the prices, while moving above and below the MA, ultimately return to it. This is a key sign for making buy/sell decisions.

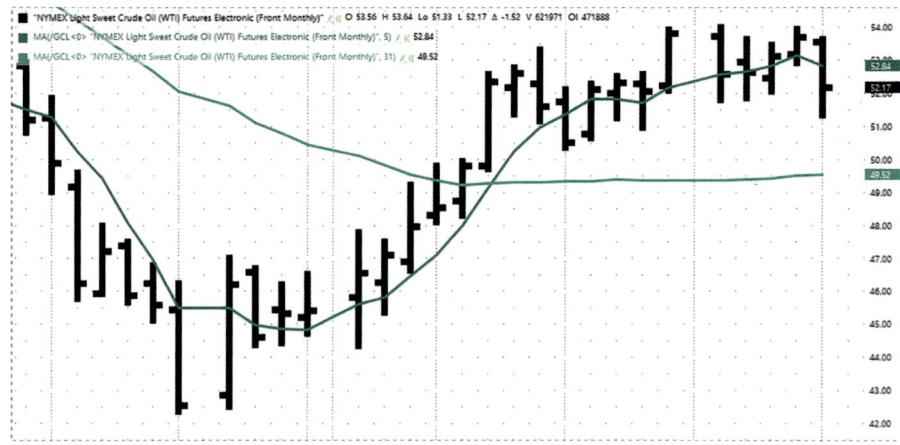

Fig. 10-7. Daily Bar Chart with 30-day & 5-day Moving Averages

3. *Bollinger bands.* Another way to monitor variances from the norm is through the use of Bollinger bands. Essentially, these chart lines "band" the variances both above and below the current trending norm. They indicate a variance level of two standard deviations from the norm. With these, it is easy to see when prices have left the upper

and lower ranges of deviation. Again, when these conditions occur, there is a high probability that the prices will return back to levels inside these boundaries.

4. *Relative strength index.* Relative strength index (RSI) is a momentum oscillator that measures the speed and change of price movements. RSI oscillates between 0 and 100. Traditionally, RSI is considered *overbought* when above 70 and *oversold* when below 30. RSI can also be used to identify the general trend. Understanding the exact RSI calculation is not necessary in order to understand how to use this indicator. Figure 10-8 is a daily bar chart with Volume, 5-, 10- and, 20-day MAs, Bollinger Bands and, the RSI study. Note that the current RSI is slightly above "50" or, "neutral".

Fig. 10-8. Daily Bar Chart with Volume, 5-, 10- and, 20-day MAs, Bollinger Bands & RSI

Price signals

As with trend analysis and market indicators, there are several types of price signals. We will deal with a few of the ones that are more common and easy to use.

1. *Support and Resistance.* As prices move up and down, traders make decisions as to when to continue to buy in an uptrend and when to sell in a downtrend. In other words, they try to determine when the current trend will exhaust itself and change direction. One way to do this is to look at the support and resistance price levels. *Support* represents a price level at which buyers will step back into the market after a period

of selling. This interest establishes a floor price. Traders find value at this level and start to buy up the contracts again. In some cases, traders who have been selling contracts during the downtrend may be buying them back to take some profits. *Resistance* is the price level at which the market is no longer interested in buying contracts. The price is deemed to be too high and sellers reenter the market, thus establishing a ceiling price.

So how do we establish these pricing points? When one trend line connects two or more price points and a second trend line connects two or more price points in parallel fashion, they form a *channel*, as shown in Figure 10–9. Channels have significance in that traders look for prices to move above or below the confines of the channel. This is referred to as a *breakout*, and depending on the number of days that form the channel, this can occur with good momentum, resulting in a large price move in that direction.

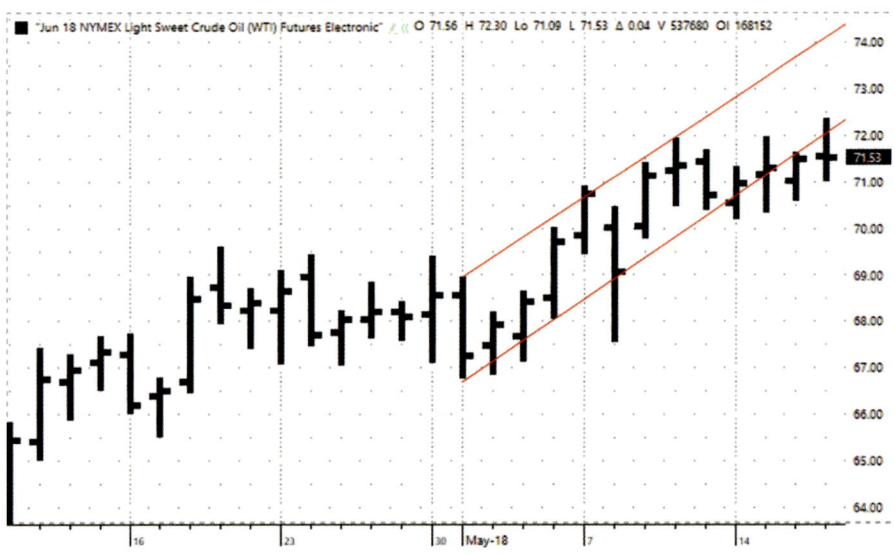

Fig. 10–9. A channel chart: A trend line connecting two or more price points with a second trend line connecting price points in parallel fashion.

Figure 10-10 below shows when we draw upper and lower trend lines, the lines continue through price points on the right vertical axis. Where the upper trend line crosses the right axis is the *Resistance point*. The price where the lower trend line crosses the right axis is the *Support point*. Theoretically, then, these represent both the maximum the market is willing to pay as well as the minimum at which is it willing to sell.

This chart indicates that resistance is about $72.50 and support is about $71.20. Traders will then look to see if prices trade above or below these levels. If they do, there will be a flurry of activity in the direction of the move.

2. *Tops and bottoms.* Since we are on the subject of support and resistance, we can discuss price signals related to those concepts. As we have said, traders are interested solely in price movement. And, support and resistance levels represent buying and selling interest. But what happens when the buyers or sellers step in to halt the moves higher or lower? They are testing the points of support and resistance. If the sellers cannot break through support, it is a result of buyers stepping in. As mentioned above, that sets a floor or bottom price on that day. Likewise, if buyers test the resistance price and sellers step in to prevent a breach of that level, a ceiling or top is established.

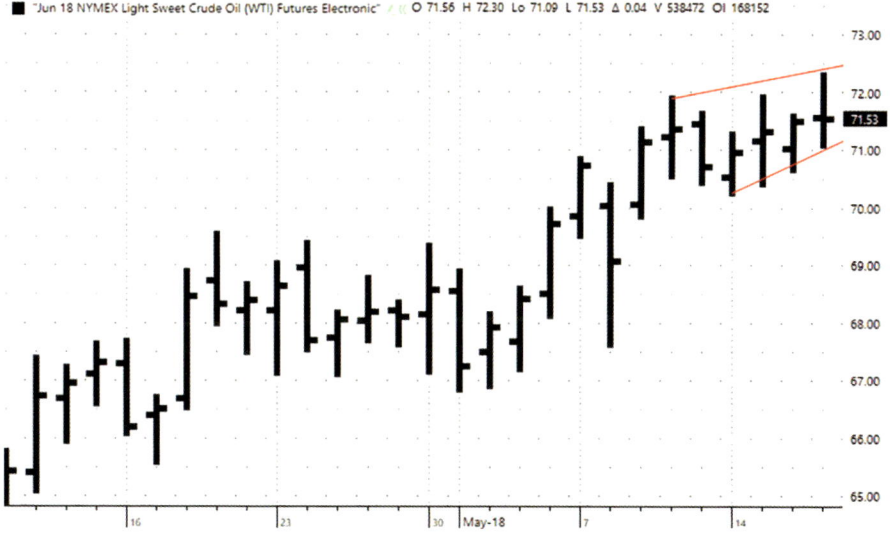

Fig. 10-10. Support & Resistance Levels

While a one-day occurrence of these events is not a very strong indicator of a change in direction, the more a bottom or top is tested and holds, the more significant that price level becomes. Let's consider an example of this with crude oil traders. In this instance, the crude oil traders are trying to sell June contracts and push the price down to the $71.20 support level on the chart above. Buyers step in at that price, and the sell-off fails. The next day, sellers again attempt to push prices down to $71.20, and again the move fails. The market now begins to see $71.20 as a stronger support price. We refer to this as a *double bottom*. While this is still a good indicator of price levels, a third day, or *triple bottom*, is a very strong indicator that prices will rally higher. Traders have no choice but to recognize the buying interest at $71.20 and thus will buy contracts until the resistance or top is tested. The same holds true for resistance levels, but in reverse. The more tops are established, the stronger the level at which sellers will step in and sell contracts.

3. *Head-and-shoulders reversal patterns.* These are identifiable, three-day price patterns that signal a change in direction and can be used for long-term or short-term trend analysis. This consists of three consecutive trading days where the middle day's high or low is higher or lower than that of the other two days. The first day then represents the *left shoulder.* The second day is the *head,* and the third day is the *right shoulder.* Looking at Figure 10-11 below, without all the trend lines, we can see that on January 15th, the high for the day was higher than the previous day, the 14th. We are now looking for the completion of the pattern the next day. And on January 16th, the high for the day was lower than the prior day, the 15th. Now you can see the pattern whereby January 14th is the left shoulder, January 15th value is the head, and January 16th is the right shoulder. The right shoulder "leans" in the direction of the price change. In this case, prices reversed from an uptrend to a downtrend. There are also reverse head-and-shoulders patterns. These occur in an upside-down fashion and signal a move from a downtrend to an uptrend. Looking at January 23rd,

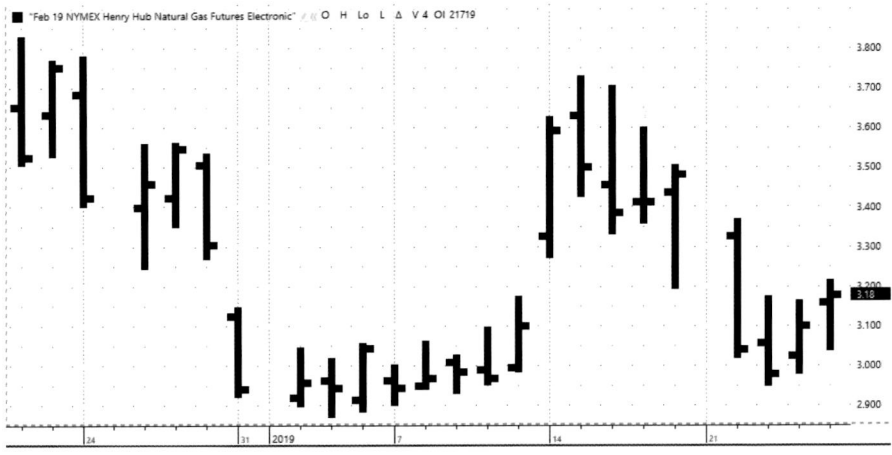

Fig. 10-11. Head & Shoulders Reversal Patterns

we see that the day's low was lower than that on January 22nd. Then on January 24th, the pattern was completed as that day's low was higher than that on January 23rd. Since the right shoulder is leaning upward, the trend is now upward.

4. *Consolidation patterns.* When upper and lower trend lines are drawn and are parallel to one another and perpendicular to the Y axis, they form a rectangular shape. The upper trend line does represent Resistance, with the lower trend line indicating Support.

In this pattern, prices will move up and down within the rectangle. This consolidation is indicative of market indecision. Traders are not really sure what direction prices should take and so, prices are "range-bound". It is a battle between buyers and sellers. The key here is the number of days this pattern continues to exist. The longer traders battle, the more momentum builds up for when prices break out of this range. Think of it as a spring that winds tighter and tighter for each day prices stay within the consolidation range. That means a very large price movement will occur in the direction of the breakout.

Summary Points: Technical Analysis

1. Technical analysis is mainly used by day traders, or speculators, to determine the greater probability of one thing happening over another (price direction).
2. By plotting prices, technical charts actually record the behavior of the market participants. Technical analysts look for these patterns to repeat.
3. There are three main types of technical charts used:
   - *Daily bar chart.* Shows the open/high/low/close prices.
   - *Close-only* or *line chart.* Shows the settlement price for each time segment.
   - *Candlestick chart.* Shows the same information as the daily bar chart but also indicates up and down days.
4. Trend lines are used to identify current and past trends but must touch on at least two pricing points to have significance.
5. Examples of simple trend indicators include the following:
   - *Volume.* This illustrates the amount of activity behind the price movement, which either reinforces it or fails to support it.
   - *Moving averages.* Traders look to the statistical regression to the mean as a predictor of price direction.
   - *Relative strength index.* A momentum indicator that identifies both the speed and change in price with a resultant overbought, oversold, or neutral market condition.
6. Examples of price signals include the following:
   - *Support.* The price level at which buyers will step in, establishing a floor.
   - *Resistance.* The price level at which sellers will step in, establishing a ceiling.
   - *Tops and bottoms.* These are recurring highs or lows that hold, thus establishing strong support or resistance. Double or triple tops and/or bottoms are very strong indicators of a possible change in price direction.
   - *Head-and-shoulders reversal pattern.* A three-day price pattern whereby the middle day's high or low is higher or lower than the other two, thus

forming a left shoulder, head, right shoulder configuration. The right shoulder leans in the direction of the price change.
- *Consolidation.* This represents several days of trading stuck within a certain high/low range. It indicates price indecision by the market and a battle between buyers and sellers. The more days within the pattern, the greater the velocity of any break out of the pattern.

Review Questions

1. _____ Analysis makes use of charts to understand price movements in the marketplace.
2. First, we must identify the _____ trend and then, determine if this trend will continue or, change direction.
3. Name the (3) most popular types of technical charts.
    a. _____
    b. _____
    c. _____
4. The price point at which buying stops and sellers enter the market is referred to as, "_____".
5. The price point at which selling stops and buyers enter the market is referred to as, "_____".
6. A "close only" or, "Line" chart shows what information? _____
_____
7. A Daily Bar chart shows what information?
    a. _____
    b. _____
    c. _____
    d. _____
8. What's another name for a Daily Bar chart? _____
9. A candlestick chart shows what information?
    a. _____
    b. _____
    c. _____
    d. _____ and,
    e. _____
10. A simple, 3-day price reversal signal is the _____ _____ pattern.

Notes

1. Martin J. Pring, *Martin Pring's Introduction to Technical Analysis*, 2nd ed. (New York: McGraw Hill, 2015), 13.
2. Ibid.
3. Ibid., 14.

11

Risk Controls in Energy Commodity Trading and Hedging

Overview

On December 2, 2001, Enron Corp., at the time the world's largest energy trading company, declared bankruptcy, causing a loss of $11 billion for its shareholders and billions more for its trading counterparties. At the time, it was the largest bankruptcy filing in US history. As events unfolded and the investigations took place, it was revealed that there were several off-sheet "paper" companies churning out false earnings. These were "mark-to-market", *unrealized* earnings that had no cash gains associated with them. Ultimately, a lack of controls, or a failure to adhere to them, allowed this to occur. Top executives at Enron were convicted and sent to prison, and their outside auditors, Arthur Andersen, went out of business.

In this chapter, we will learn about other famous cases where financial disasters took place due to a lack of controls and oversight. We will explore concepts such as mark-to-market and value at risk (VaR), financial risk measures that are mandatory for today's publicly-traded energy companies that deal in financial derivatives.

Key Learning Points: Risk Controls in Energy Commodity Trading and Hedging

- Severe losses by "rogue" traders led to the establishment of controls for financial derivative trading in the banking and finance businesses.
- There were common elements among these trading disasters.
- Risk protection measures were later made mandatory for the energy industry.
- Companies face more than just financial risk. There is also legal, operational, and credit risk.
- Necessary risk controls, measures, reports, and organizational structures need to be established.
- Mark-to-market gives the current value of all open trading positions based on daily market prices.

- *P&L* is the estimated profit and/or loss determined by the mark-to-market calculations.
- *Value at risk (VaR)* is a theoretical measure of the maximum potential loss for a trading book.
- Corporations face various risk exposures. Among the risk types are the following:
  - Financial
  - Market
  - Counterparty
  - Operation
  - Credit
  - Legal
- Publicly traded energy companies engaged in trading financial derivatives were required to implement risk controls by FY2000.
- Companies need to have defined risk control structures in place, including:
  - Standard risk metrics
  - Daily reporting requirements
  - Risk policies and procedures
  - Violations reporting
  - Independent risk control staff headed by a chief risk officer
  - Risk oversight committees comprised of top executives

The following case studies are brief descriptions of catastrophic losses by traders with little or no oversight. They are reproduced from Pennsylvania State University's course, "EBF 301: Global Finance for the Earth, Energy, and Materials Industries," authored by Farid Tayari and Tom Seng.

Case Study 1: Barings Bank, PLC[1]

In February 1995, Nick Leeson, a "rogue" trader for Barings Bank, UK, single-handedly caused the financial collapse of a bank that had been in existence for hundreds of years. In fact, Barings had financed the Louisiana Purchase between the US and France in 1803. Leeson was dealing in risky financial derivatives in the Singapore office of Barings. He was the lone trader there and was betting heavily on options for both the Singapore (SIPEX) and Nikkei exchange indexes. These are similar to the Dow Jones Industrial Average (DJIA) and the S&P 500 indexes here in the US.

In the early 90s, Barings decided to get into the expanding futures/options business in Asia. They established a Tokyo office to begin trading on the Tokyo Exchange. Later, they would look to open a Singapore office for trading on the SIMEX. Leeson requested to set-up the accounting and settlement functions there and direct trading floor *operations* (different from actual trading). The London office granted his request, and he went to Singapore in April 1992. Initially, he could only execute trades on behalf of clients and the Tokyo office for "arbitrage" purposes. After a good deal of success in this area, he was allowed to pursue an official trading license on the SIMEX. He was then given some "discretion" in his executions, meaning, he could place orders on his own (speculative, or "proprietary" trading).

Even after given the right to trade, Leeson still supervised accounting and settlements. And there was no direct oversight of his "book" and he even set up a "dummy" account in which to funnel losing trades. So, as far as the London office of Barings was concerned, he was always making money because they never saw the losses and rarely questioned his request for funds to cover his "margin calls"… He took on huge positions as the market seemed to "go his way." He also "wrote" options, taking on huge risk.

He was, in fact, perpetuating a "hoax" in his record-keeping to hide losses. *He would set the prices put into the accounting system and "cross-trade" between the legitimate, internal, accounts and his fictitious "88888" account.* He would also record trades that were never executed on the Exchange.

In January 1995, a huge earthquake hit Japan, sending its financial markets reeling. The Nikkei crashed, which adversely affected Leeson's position (remember, he had been selling Options). It was only then that he tried to hedge his positions, but it was too late. By late February, he faxed a letter of resignation, and when his position was discovered, he had lost ($1.4 billion USD). Barings, the bank which financed the Louisiana Purchase between the US and France, became insolvent and was sold to a competing bank for $1.00! (If you are interested in more details regarding this infamous case, you can read *Rogue Trader* by Nick Leeson himself. There is also a movie of the same name starring Ewan McGregor….)

Case Study 2: Orange County, CA[2]

Robert Citron was the Treasurer for Orange County, California, in the early 90s. He was solely responsible for investing several of the county's funds which totaled about $7.5 billion USD. Despite having no background in trading financial instruments, he decided to invest in risky interest rate swaps that were tied to the US Treasury Department's rates.

Citron was a County Tax Collector with no college degree who was later elected to the position of Orange County Treasurer. In this capacity, he was able to push for California legislative approval for county treasurers to increase their use of financial instruments for investment and fund management.

He was attempting to arbitrage the difference between short-term and long-term interest rates. His position was sound and he could make money so long as short-term rates remained low. During his tenure, the average return on county investments was a healthy 9.4%, but interest rates had been low for that long. The position he took would lose money if interest rates rose. And, he inflated the county's volumetric position by entering into other derivatives that would also be negatively impacted by higher interest rates.

Beginning in February 1994, the Federal Reserve Board made the first of six consecutive interest rate hikes. Between February and May of that year, the County had to produce $515 million in cash (margin) to cover its position. Further margin calls would occur throughout year, leaving the County's cash reserves at only $350 million by November, 1994. When word got out about the County's troubles raising cash, investors sought to retrieve their money, and by December 6, 1994, the County declared bankruptcy and lost ($1.64) billion.

Case Study 3: Metallgesellschaft AG (MG)[3]

MG was a huge, German industrial conglomerate that decided to open an energy trading office in the US in the early 90s.

The original plan was threefold:
- Sell refined products in the forward physical market.
- Invest in refining capacity to produce the products.
- Hedge the forward sales through financial derivatives.

When the strategy was first implemented in 1992, current physical prices were lower than the futures prices. So the sales contracts were set at those higher prices. And it meant that purchasing the "near" month futures contracts would be profitable. So MG developed a strategy whereby they would cover the long-term, fixed-price sales by buying contracts in these few, near months. This is known as a "stack-and-roll" approach. As each month "rolled off," they would merely buy contracts in the next month. It was their intent to continue this process until the physical product sales contracts expired in (10) years. This strategy worked as long as the futures market was "backwardated," whereby each successive month is less than the prior one [since they had to buy the contracts].

One of the major flaws in this approach, however, was the volume of contracts being traded since they were "loading up" on closer month contracts. Add to that the fact that they would not get paid for the product sales for years out, and you begin to have a cash flow problem where margin calls are concerned. Their position in the Fall of 1993 was estimated to be between 160 to 180 million barrels stretched-out over the following (10) years.

In 1993, prices fell as the market received a "bearish" signal from OPEC on production quotas. This lowered futures prices and reversed the market from "backwardated" to "contango," whereby each successive month's price is higher than the prior one. Faced with this position [huge margin calls], MG management was changed, and the new team was directed to close all positions. This resulted in losses on the futures purchases totaling almost ($1.5) billion USD. They had to seek bailout funds from one of their banks, and in return, had to sell-off several divisions. Today, the German industrial giant no longer exists having been bought-out by a competitor.

Key Lessons Learned by Examining the Case Studies

There were some common factors in each of these cases:
- There was a single rogue trader or small groups of rogue traders and little supervision over the traders' decision-making process.
- There was risky use of financial derivatives.
- There was a lack of real accounting/auditing oversight and/or the trader(s) controlled these.
- There were no trading policies, controls, etc. in place.
- Trade losses were hidden.
- There was a lack of executive knowledge and understanding of the inherent risks in trading.
- Trading positions were increased to lessen impact of losses, leading to increased exposure (so-called doubling-down).
- These events, along with others, prompted the financial industry to institute ways to monitor, track, and stay on top of financial derivative trading. These same methods would later have to be adopted by publicly traded energy companies in the United States.

Risk Policies and Controls for Energy Commodity Derivatives

The use of risk controls for those who engage in energy financial derivatives has probably never been as important as it is today. As delineated earlier, corporations face myriad risks, not the least of which is financial exposure to volatile energy markets. We have also covered the vast number of factors that can turn energy prices on a dime. And as previously indicated, the involvement of supercomputers in trading, along with an almost 24/7/365 ability to trade, market participants today can lose money more quickly than ever.

The downfall of Enron in late 2001, and the subsequent exit of the country's top energy marketing companies, led to a loss in counterparty liquidity for the industry. Banks and financial institutions entered the game only to collapse in 2008, again exacerbating the financial counterparty liquidity situation. Online exchanges such as ICE and ClearPort, however, provided clearing mechanisms to support a healthy OTC market. NYMEX gained importance for those seeking the protections of a formal futures exchange for energy derivatives.

In the late 1990s, both the SEC and the CFTC ruled that, effective with their FY2001 annual filings, all publicly-traded energy companies must adopt risk controls and report in earnings the value of their outstanding (open) derivative positions. (This addressed the "license to steal" that Enron used to falsify earnings by creating shell companies that had earnings on paper in transactions with each other. An executive at Enron, Andy Fastow, set the daily market values used for mark-to-market gains and losses.[4])

In a post-Enron energy trading world, having speculative financial derivative positions was frowned upon by Wall Street analysts. Those energy trading firms that survived the losses suffered by dealing with Enron immediately changed their trading strategies. The most notable change was merely a switch from energy trading companies to energy services, thus removing the negative moniker. And Wall Street was now much more interested in a company's open derivatives positions, since it was definitely a measure of how much risk they were taking on.

So what makes a good risk control program? It starts with a solid risk policy. That policy needs to be clear on the purpose for which the company is entering into energy financial derivative contracts. For most companies these days, it is to hedge against price and market risk that could put their earnings at risk.

The policy also needs to lay out the specifics of the following:

- *Risk oversight committee.*[5] The risk oversight committee (ROC) is comprised of corporate-level executives and a chief risk officer. While the marketing and/or trading divisions should attend committee meetings, they should not be voting members to keep the decisions of the committee "at arm's length."
- *Risk control desk.* Reports to chief risk officer; corporate level.
  - Director/manager
  - Risk analysts
    ◊ Confirmation analyst. Works with financial counterparties to confirm transactions and send out official confirmations.
    ◊ Risk system input analyst. Transaction input to risk management system.
    ◊ Risk control analysts. Various analytic functions and risk report compilations and verification of prices used for risk reports ("curves").
- *Risk management system.* Software product to run daily, monthly, quarterly, and annual risk reports, including mark-to-market, value at risk (VaR), stress testing, etc.
- *Risk control procedures*

- Detailed, SOX-compliant (Sarbanes-Oxley Act of 2002) descriptions of the functions of each position within the risk control area.
- Detailed, SOX-compliant descriptions of the various risk reports to be run, along with the specific steps taken to produce them.
- Detailed, SOX-compliant descriptions of the frequency of generating the risk reports and the distribution lists for them.
- Risk reports should be generated at the individual trader level all the way up to encompassing the entire book.
- Risk control compliance measures
  - Key risk measures
    - Mark-to-market. The value of all open positions marked against the daily market settlement prices for each derivative. These represent *unrealized* gains/losses, since the positions have not been closed. They are an estimate of the actual gain/loss should the positions be closed out at the daily market prices.
    - Value at risk. Theoretical maximum loss on total book for a given period of time, at a given confidence level, defined holding period, and at expected market conditions.
      - Expressed as a single value, e.g., "$10-million-dollar loss at 98% confidence level."
      - Historical prices and Monte Carlo simulation (random number generator) are used to establish several possible mark-to-market scenarios.
    - P&L. Profit/loss for a given time period comprising real gains/losses as well as mark-to-market gains/losses.
  - All trader/marketers with the authority to execute/trade derivatives on behalf of the company should be given specific mark-to-market and VaR limits.
  - All trader/marketers with the authority to execute/trade derivatives on behalf of the company should be made to sign the risk control policy. This should include those with the authority to execute physical fixed-price transactions, as those also contain price risk.
  - Specific penalties for noncompliance need to be spelled out, including termination for violation of the risk policies. One particular problem that can exist is that of a trader "drawering" a trade. That is, they have executed a trade that is now losing money but do not report it to the risk control desk. Instead, they hide the trade, hoping the position will change. The term *drawering* comes from the act of putting the associated deal sheet in a desk drawer rather than submitting it to the risk control desk.

- Limitations need to be established on such variables as volume, mark-to-market losses, and value at risk. These need to be set at both the individual trader level and for the entire book.
- Policy limit violations. Once a limit set by the risk oversight committee has been breached, either by an individual trader or for the entire book, the ROC should be immediately notified. Actions to be taken to correct the risk violation and bring it back into compliance should be decided by the ROC and implemented immediately. This could involve liquidating open positions or terminating a trader whose actions have added undue risk for the company.

Summary Points

1. Corporations face various risk exposures. Among the risk types are the following:
   a. Financial
   b. Market
   c. Counterparty
   d. Operation
   e. Credit
   f. Legal
2. Severe losses by "rogue" traders led to the establishment of controls for financial derivative trading in the banking and finance businesses.
3. Publicly traded energy companies engaged in trading financial derivatives were required to implement risk controls by FY2000.
4. Companies need to have defined risk control structures in place, including:
   a. Standard risk metrics
   b. Daily reporting requirements
   c. Risk policies and procedures
   d. Violations reporting
   e. Independent risk control staff headed by a chief risk officer
   ○ Risk oversight committees comprised of top executives
5. *Mark-to-market* gives the current value of all open trading positions based on daily market prices.
6. *P&L* is the estimated profit and/or loss determined by the mark-to-market calculations.
7. *Value at risk (VaR)* is a theoretical measure of the maximum potential loss for a trading book.

Review Exercises

1) Name (5) risks that corporations face:
   a. _____
   b. _____
   c. _____
   d. _____
   e. _____

2) Name (3) standard risk control measures:
   a. _____
   b. _____
   c. _____

3) What elements should a solid risk policy contain?
   a. _____
   b. _____
   c. _____
   d. _____

4) Why do companies "hedge"? _____

5) What is "mark-to-market"? _____

6) What is "VaR"? _____

7) What does "P&L" mean? _____

8) The internal governing body which oversees financial energy trading is known as the _____

9) "SOX" stands for _____

10) Name (5s) common elements among the risk case studies presented:
    a. _____
    b. _____
    c. _____
    d. _____
    e. _____

Notes

1. Farid Tayari and Tom Seng, "Case Study 1: Barings Bank PLC," from *EBF 301: Global Finance for the Earth, Energy, and Materials Industries*, College of Earth and Mineral Sciences, Pennsylvania State University, https://www.e-education.psu.edu/ebf301/node/569.
2. Farid Tayari and Tom Seng, "Case Study 2: Orange County, CA," from *EBF 301: Global Finance for the Earth, Energy, and Materials Industries*, College of Earth and Mineral Sciences, Pennsylvania State University, https://www.e-education.psu.edu/ebf301/node/570.
3. Farid Tayari and Tom Seng, "Case Study 3: Metallgesellschaft (MG)," from *EBF 301: Global Finance for the Earth, Energy, and Materials Industries*, College of Earth and Mineral Sciences, Pennsylvania State University, https://www.e-education.psu.edu/ebf301/node/571.
4. For a good description of what took Enron down, see Robert Bryce's book, *Pipe Dreams: Greed, Ego, and the Death of Enron*. I prefer this to *Smartest Guys in the Room* since Bryce interviewed more than 300 ex-Enron employees in his research. In addition to the scheme perpetrated by Ken Skilling and Fastow, he unveils a lot of the scandals that were plaguing Enron at the time which make for an interesting read.
5. Risk committees take on numerous names, such as "Risk Control Steering Committee," "Risk Control & Hedge Committee," etc.

Appendix A

EIA's "Weekly Petroleum Status Report: Highlights" for Week Ending January 25, 2019[1]

U S EIA, "Summary of Weekly Petroleum Data for the week ending January 25, 2019."

U.S. crude oil refinery inputs averaged 16.5 million barrels per day during the week ending January 25, 2019, which was 586,000 barrels per day less than the previous week's average. Refineries operated at 90.1% of their operable capacity last week. Gasoline production increased last week, averaging 9.9 million barrels per day. Distillate fuel production decreased last week, averaging 5.0 million barrels per day. U.S. crude oil imports averaged 7.1 million barrels per day last week, down by 1,108,000 barrels per day from the previous week. Over the past four weeks, crude oil imports averaged about 7.7 million barrels per day, 4.5% less than the same four-week period last year. Total motor gasoline imports (including both finished gasoline and gasoline blending components) last week averaged 523,000 barrels per day, and distillate fuel imports averaged 135,000 barrels per day. U.S. commercial crude oil inventories (excluding those in the Strategic Petroleum Reserve) increased by 0.9 million barrels from the previous week. At 445.9 million barrels, U.S. crude oil inventories are about 7% above the five year average for this time of year. Total motor gasoline inventories decreased by 2.2 million barrels last week and are about 5% above the five year average for this time of year. Finished gasoline and blending components inventories both decreased last week. Distillate fuel inventories decreased by 1.1 million barrels last week and are about 2% below the five year average for this time of year. Propane/propylene inventories decreased by 3.6 million barrels last week and are about 2% above the five year average for this time of year. Total commercial petroleum inventories decreased last week by 4.8 million barrels last week. Total products supplied over the last four-week period averaged 20.7 million barrels per day, down by 0.2% from the same period last year. Over the past four weeks, motor gasoline product supplied averaged 8.9 million barrels per day, up by 1.4% from the same period last year. Distillate fuel product supplied averaged 4.0 million barrels per day over the past four weeks, down by 3.1% from the same period last year. Jet fuel product supplied was down 0.9% compared with the same four-week period last year.

Note

1. EIA, "Summary of Weekly Petroleum Data for week ending January 25, 2019", http://ir.eia.gov/wpsr/wpsrsummary.pdf

Appendix B

EIA's "Weekly Natural Gas Storage Report" for Week Ending January 25, 2019[1]

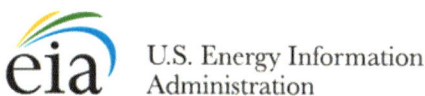

‹ SEE ALL NATURAL GAS REPORTS

Weekly Natural Gas Storage Report

for week ending January 25, 2019 | Released: January 31, 2019 at 10:30 a.m. | Next Release: February 7, 2019

Working gas in underground storage, Lower 48 states   Summary text CSV JSN

	Stocks billion cubic feet (Bcf)				Historical Comparisons			
					Year ago (01/25/18)		5-year average (2014-18)	
Region	01/25/19	01/18/19	net change	implied flow	Bcf	% change	Bcf	% change
East	527	566	-39	-39	529	-0.4	575	-8.3
Midwest	606	673	-67	-67	601	0.8	663	-8.6
Mountain	114	121	-7	-7	138	-17.4	149	-23.5
Pacific	178	185	-7	-7	222	-19.8	242	-26.4
South Central	771	823	-52	-52	720	7.1	896	-14.0
Salt	278	295	-17	-17	166	67.5	248	12.1
Nonsalt	493	528	-35	-35	554	-11.0	648	-23.9
Total	2,197	2,370	-173	-173	2,211	-0.6	2,525	-13.0

Totals may not equal sum of components because of independent rounding.

Summary

Working gas in storage was 2,197 Bcf as of Friday, January 25, 2019, according to EIA estimates. This represents a net decrease of 173 Bcf from the previous week. Stocks were 14 Bcf less than last year at this time and 328 Bcf below the five-year average of 2,525 Bcf. At 2,197 Bcf, total working gas is within the five-year historical range.

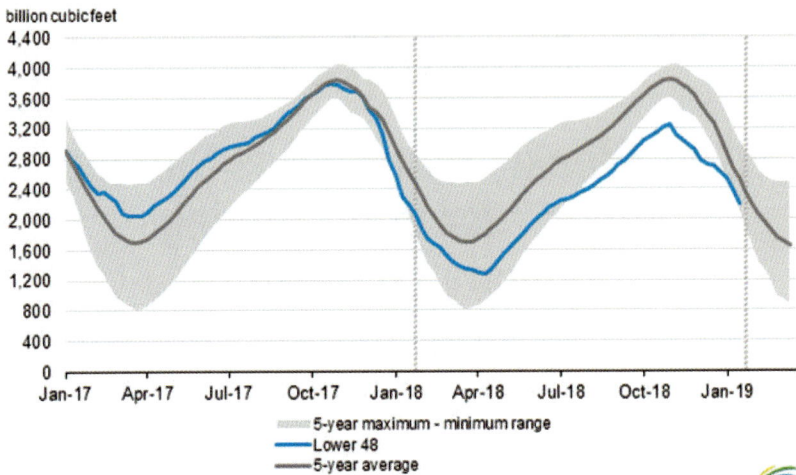

Note

1. US EIA, "Weekly Natural Gas Storage Report," January 31, 2019, http://ir.eia.gov/ngs/ngs.html

Appendix C
Energy Markets Risk Management Glossary

Global Association of Risk Professionals (GARP)

A

American Style Option
An option which can be exercised at any time before its expiry date.

Ancillary Services
Any service required by a system operator to deliver electricity to the ultimate consumer. Ancillary services can include regulation, spinning reserves, non-spinning reserves, and replacement reserves.

Anthracite Coal
The highest rank of black coal with primary use in residential and commercial heating.

API
American Petroleum Institute. The API publishes weekly information on U.S. petroleum stock figures, refinery throughput, imports, exports, and stock levels. The API also established the system for grading crude oils by gravity *(see API Gravity)*.

API gravity
A scale expressing the gravity or density of liquid petroleum products, devised jointly by the American Petroleum Institute and the National Bureau of Standards. The higher the API gravity, the lighter the crude.

Arbitrage
The simultaneous purchase of a commodity/derivative in one market and the sale of the same, or similar, commodity/derivative in another market in order to exploit price differentials. In financial markets, the practice of making a profit from temporary price differentials between two or more markets for the same asset.

Asian Option (Average Price Option)
An option that is exercised against an average over a period of time.

At-The-Money Option (ATM)
An option whose exercise price is equal, or close to, the current price in the underlying market.

B

Backwardation
A market where the price for nearby delivery is higher than for further forward months. (The opposite of backwardation is contango).

Balancing
The requirement imposed by power grids or natural gas pipelines to ensure supply and demand be equal over a certain period of time.

Barge
Vessel carrying oil, usually on rivers, containing between 8,000 and 50,000 barrels.

Barrel (bbl)
A volumetric unit of measure for crude oil and petroleum products. One barrel is equal to 42 US gallons.

Baseload
The generation capacity that meets the minimum amount of demand throughout the entire 24 hour day (around-the-clock demand). Baseload power is generally supplied from plants that cannot be ramped up/down as quickly as peaking generation plants. Since Baseload demand is generally predictable and steady, it's less expensive than peak power.

Baseload Plant
A plant normally operated to take all or part of the minimum continuous load of a system, and which consequently produces electricity at an essentially constant rate. A base load plant typically has relatively high fixed costs and low unit operating costs. Traditionally, nuclear plants have been considered as base load plants.

Basis
The price gap for the same physical commodity, but at two different geographic locations.

Basis Swap
Basis swaps are used to hedge exposure to basis risks, such as locational risk or time exposure risk.

Bearish
The belief that the commodity price is going to fall.

Bid
An indication of willingness to buy a specified amount of a commodity at a specific price.

Bid/Ask
A measure of market liquidity, also known as bid/offer. The bid is the price level at which buyers are willing to buy, and the ask is the price level at which sellers are willing to sell. The tighter the spread, the higher the liquidity.

Bituminous Coal
A black or dark brown coal used as fuel in steam-electric power generation and manufacturing.

Black-Scholes
Options pricing theory derived by Fisher Black and Myron Scholes, based on the theory that price volatility is random around a given trend.

Brent
The most commonly traded North Sea crude oil. Brent has an API of about 37.5

British Thermal Unit (btu)
The quantity of heat required to raise one pound of water one degree Fahrenheit at or near its point of maximum density. An MMbtu (-mil Btus) is roughly equivalent to a Mcf (a thousand cubic feet).

Bullish
The belief that the commodity price is going to rise. Opposite of bearish.

Burner Tip
The point at which natural gas is used as a fuel.

C

Call Option
An option that gives the buyer (holder) the right but not the obligation to buy a specified quantity of an underlying futures contract at a fixed price, on or before a specified date. The grantor of the option is obliged to deliver the future at the fixed price if the holder exercises the option.

Capacity
The power output rating of a generator, typically in megawatts, measured on an instantaneous basis.

Cap and Trade Approach (Emissions)
The cap and trade approach establishes an aggregate emissions allowance cap—maximum limit to emit emissions—on all sources of emissions. The polluters must limit their emissions to the established emission allowance caps. If one of the polluters emits less than the allocated emission allowance cap, the polluter can sell its emission allowance rights, subject to the established cap, to another polluter that needs to buy emissions allowances that allow them to emit more than what its cap allows.

Cash and Carry
An arbitrage transaction involving the simultaneous purchase of a cash commodity with borrowed money and the sale of the appropriate futures contract.

Cash Market
The physical market underlying a futures or options contract.

Cash Settlement
The settlement of futures or options by paying a cash difference, rather than taking/making physical delivery.

CDD
Cooling Degree Days. A measure of how warm a location is over a period of time relative to a base temperature, most commonly specified as 65 degrees Fahrenheit. The measure is computed for each day by subtracting the base temperature (65 degrees) from the average of the day's high and low temperatures, with negative values set equal to zero. Each day's cooling degree days are summed to create a cooling degree day measure for a specified reference period. Cooling degree days are used in energy analysis as an indicator of air conditioning energy requirements or use.

CFTC
Commodity Futures Trading Commission

CIF
Cost, Insurance and Freight. Shipping Incoterm refers to cargos for which the seller pays for the transportation and insurance up to the port of destination.

City-Gate
Physical location where gas is delivered by a pipeline to a local distribution company.

Clearing
The process of matching trades, settling trades and provision of a guarantee for traded contracts. Often a service performed by exchanges.

Clearing Fee
A fee charged by a clearing house for clearing trades.

Close Out
Finalizing a transaction by making an equal and opposite trade to an open position.

CME
Chicago Mercantile Exchange

CNG
Highly compressed natural gas, commonly used as an automotive fuel.

Cogeneration
The production of both electricity and useful thermal energy from the same energy source. Natural gas tends to be a favored fuel for combined-cycle cogeneration units, in which waste heat is converted to electricity.

Combination Hedging
A risk management strategy that uses a combination of hedges using different derivative instruments.

Combined Cycle
The combination of one or more gas turbine and steam turbines in an electric generation plant. An electric generating technology in which electricity is produced from otherwise lost waste heat exiting from one or more gas (combustion) turbines. The heat is routed to a conventional boiler or to a heat recovery steam generator for use by a steam turbine in the production of electricity. This process increases the efficiency of the electric generating unit.

Condensate
Very high API crude oil, which in its natural state is in gaseous form, but condenses to liquid upon production.

Congestion
A condition that occurs when insufficient transfer capacity is available to implement all of the preferred schedules for electricity transmission simultaneously.

Contango
Market situation where prices are higher for forward delivery dates than for nearer delivery dates. (Opposite of backwardation).

Conway
Conway, Kansas. The main propane trading hub in the Midwest United States.

Cracking
Refining process to break large molecules into smaller ones.
Crack Spread
The spread differential which represents refining margins.
Credit Risk
Credit Risk is the risk of loss to a company due to non-payment of a bill or a loan, or to a nonperformance of a contract. When a company provides a service or goods and expects payment later, the company assumes the credit risk of the counterparty not paying.
Cushing
Cushing, Oklahoma. The delivery point for NYMEX WTI crude oil futures.

D

Day-Ahead Market
Forward markets where electricity quantities and market clearing prices are calculated individually for each hour of the day on the basis of participant bids for energy sales and purchases.
Day-Ahead Schedule
A schedule prepared by a scheduling coordinator or the independent system operator (ISO) before the beginning of a trading day. This schedule indicates the levels of generation and demand scheduled for each settlement period that trading day.
Day Traders
A day trader establishes positions through buying and selling commodities or financial instruments several times during one trading day, without the intention of carrying these positions over to subsequent trading days. Most, if not all, of the positions established during the trading day are closed out before the end of the day's trading.
Degree Days
Degree days are measured as the number of degrees above or below a standardized temperature on any given day. In winter, traders track heating degree days week by week, or month by month, normally against a standard temperature of 65F, on the basis of how many degrees of heat are required to bring the temperature up to 65F. In summer, the market tracks cooling degree days, which are computed in the opposite manner. *(See HDD, CDD.)*
Delivered Cost
The cost of fuel, including the invoice price of fuel, transportation charges, taxes, commissions, insurance, and expenses associated with leased or owned equipment used to transport the fuel.
Delta
The rate of change of the value of an option with respect to changes in the price of the underlying commodity.
Delta Hedging
The process whereby the grantor of an option decides to buy or sell more/less of an underlying futures contract in order to protect against being declared upon by the options holder. If delta hedging, the grantor of a call option will buy more of the futures contract if it rises in value towards the strike price (as the probability of being declared

upon rises towards 100%). The grantor of a put option will typically sell more of the underlying futures contract if it slides in value (as the probability of being declared upon rises towards 100%).

Delta Neutral
A state where the grantor of an option has balanced the probability of being declared upon through buying/selling the underlying futures contract.

Demand, Oil
The rate of consumption of refined products, normally measured in millions of barrels per day.

Demand, Power
The rate at which power is delivered to or by a system at a given instant or averaged over a designated period, usually expressed in kilowatts or megawatts.

Demand Response Programs
Demand response programs are incentive-based programs that encourage electric power customers to temporarily reduce their demand for power at certain times in exchange for a reduction in their electricity bills. Some demand response programs allow electric power system operators to directly reduce load, while in others, customers retain control. Customer-controlled reductions in demand may involve actions such as curtailing load, operating onsite generation, or shifting electricity use to another time period.

Demand Side Management
All activities or programs undertaken by an electricity system or consumers to influence the amount and timing of electricity use.

Demurrage
The detention or delay of a vessel in loading or unloading beyond the time agreed upon. Demurrage charges are usually incurred for any delay.

Differential
The difference between two prices. A large and increasing percentage of all oil transactions are effected on the basis of differentials, also known as spreads, rather than outright flat prices.

DOE
United States Department of Energy

Downstream
The part of the energy cycle focused on refining crude oil, processing natural gas, as well as the selling and distribution of natural gas and refined products.

Dry Gas
Natural gas which does not contain liquid hydrocarbons. Gas is usually priced on a dry basis.

E

EFP
Exchange of futures for physical: refers to the exchange of a futures position for a physical (swap) position.

EIA

Energy Information Administration—an independent agency within the U.S. Department of Energy that develops surveys, collects energy data, and analyzes and models energy issues. The Agency must meet the requests of Congress, other elements within the Department of Energy, Federal Energy Regulatory Commission, the Executive Branch, its own independent needs, and assist the general public, or other interest groups, without taking a policy position.

Emission

The release of gases and pollutants into the atmosphere.

Emission Allowances Rights

Emission allowance rights limit—cap—the total amount of pollution, emissions, and/or pollutants that all polluting sources can emit or release. These rights limit the polluters' emissions up to the cap.

Energy Value Chain

The process of moving from energy commodities in its raw state into energy and consists of exploration and production, transportation and storage, refining and processing, and distribution and sales.

European Option

An Option that can only be exercised on the date of expiry. These typically trade in the OTC markets.

Exchange

An organized and regulated marketplace where financial assets, commodities, and other financial products are traded between. An exchange can be physical or electronic.

Exchange-Traded

Futures or options that are traded on an exchange such as NYMEX or ICE, with standard contracts and rules.

Exercise

The procedure by which an option holder takes up the rights to the contract and is delivered a long (call) or short (put) futures position by the grantor at a fixed price.

Expiry (Expiration Date)

The last day an option may be exercised. European options can only be exercised on the expiration day. American options can be exercised anytime up to the expiration date.

F

Feedstock

Raw material used in a processing plant.

FERC

Federal Energy Regulatory Commission. A U.S. federal agency created in 1977 to regulate, among other things, interstate wholesale gas and transportation of gas and electricity at "just and reasonable" rates. Located in Washington, DC, the FERC has jurisdiction over interstate electricity sales, wholesale electric rates, hydroelectric licensing, natural gas pricing, oil pipeline rates, and gas pipeline certification. FERC is an independent regulatory agency within the Department of Energy.

Firm Gas
Natural gas sold on a continuous basis for a defined contract term.

Firm Power
Electricity capacity intended to be available at all times during the period covered by a guaranteed commitment to deliver, even under adverse conditions, but subject to force majeure interruptions. Firm power consists of either firm energy, firm capacity, or both.

Force Majeure
Denotes circumstances beyond the control of a company, which force the breaking of a contract.

Forward Contract
An agreement to make or take delivery of a commodity at a known future date, at a price agreed upon at the time the agreement is made.

Free on Board (FOB)
FOB prices exclude all insurance and freight charges. Most oil is sold either FOB (effectively priced at the loading port) or CIF (effectively priced at the delivery port).

FTR
Financial Transmission Right. FTRs allow market participants to offset potential losses related to the price risk of delivering energy to the grid.

Fuel Switching
Substituting one fuel for another based on price and availability. Large industries often have the capability of using either oil or natural gas to fuel their operation and making the switch on short notice.

Futures Contract
A standardized agreement, traded on an exchange, to make or take delivery of a commodity at a known future date, at a price agreed upon at the time the agreement is made.

G

Gallon
Generally accepted across the oil industry to refer to a U.S. Gallon. There are 42 U.S. gallons in a barrel.

Gamma
The rate of change in delta per unit change in the underlying instrument.

Gasoil
An intermediate distillate product used for diesel fuel, heating fuel and sometimes as feedstock.

Gasoline
Volatile motor fuel used in cars.

Generation
The process of producing electricity by transforming other forms of energy such as steam, heat or falling water. Also, the amount of electricity produced, expressed in kilowatt-hours (kWh) or megawatt-hours (MWh).

Generator Capacity

The maximum output, commonly expressed in megawatts (MW), that generating equipment can supply to system load.

Grid

The layout of a power transmission system on a synchronized transmission network.

H

HDD

Heating Degree Days. A measure of how cold a location is over a period of time relative to a base temperature, most commonly specified as 65 degrees Fahrenheit. The measure is computed for each day by subtracting the average of the day's high and low temperatures from the base temperature (65 degrees), with negative values set equal to zero. Each day's heating degree days are summed to create a heating degree day measure for a specified reference period. Heating degree days are used in energy analysis as an indicator of space heating energy requirements or use.

Heat Rate

A measure of efficiency of a power generating unit.

Heavy Crude Oil

Has an API gravity of less than 28 degrees. The lower the API gravity, the heavier the oil.

Hedge

The reduction of risk by covering anticipated commitments in the future through a swap, future or option contract, as a means to protect energy traders from unexpected or adverse price fluctuations.

Henry Hub

The standard delivery point for the NYMEX natural gas futures contract in the US (pipeline hub located on the Louisiana Gulf Coast) where a number of interstate and intrastate pipelines interconnect through a header system operated by Sabine Pipeline.

Historic Volatility

The change in the absolute value of a commodity or instrument over a certain period, expressed as a percentage of the lowest price recorded in that period.

Hub

A geographic location where multiple participants trade services.

Hydropower Plant

A plant in which the turbine generators are driven by falling water.

I

ICE

The Intercontinental Exchange.

IEA

International Energy Agency

Index Swap
In the natural gas market in North America, index swaps are often used to hedge against location price risk (a form of basis risk). The seller receives a fixed, or otherwise determined, price and pays the buyer the published index value for natural gas from a specified location.

Integrated Energy Company
An integrated energy company operates in all areas of the energy value chain.

Interruptible Gas
Gas sold to customers with a provision that permits curtailment or cessation of service at the discretion of the supplier.

In-The-Money Option
An option which has intrinsic value. A put option is in-the-money when its strike price is above the value of the underlying futures contract. A call option is in-the-money when its strike price is below the value of the underlying futures contract.

IOC
International Oil Company.

IPP
Independent power producer—Unregulated power generators which, unlike utilities, have no franchised retail service territories. Even a plant built by an investorowned utility to serve its native retail load is not an IPP. It's still a utility plant. Also, utilities that form affiliates and build outside of their territories can be IPPs.

ISO
Independent System Operator. An independent, federally regulated entity established to coordinate regional transmission in a non-discriminatory manner and ensure the safety and reliability of the electric system.

K

Kilowatt
One kilowatt equals 1,000 watts. Abbreviates to kW.

Kilowatt-Hour (KWh)
The basic unit for pricing electric energy, equal to one kilowatt of power supplied continuously for one hour (or the amount of electricity needed to light ten 100-watt light bulbs for one hour). One kWh equals 1,000 watt-hours. One kWh = 3.306 cu ft of natural gas.

L

Landed Cost, Crude Oil
The price of crude oil at the port of discharge, includ_ ing charges associated with purchasing, transporting, and insuring a cargo from the purchase point to the port of discharge. The cost does not include charges incurred at the discharge port (e.g., import tariffs or fees, wharfage charges, and demurrage).

Lifting
The act of loading petroleum or petroleum products at a terminal or transfer point.

Glossary

Light Crude Oil

Has an API gravity higher than 33 degrees. The higher the API gravity, the lighter the crude oil.

Light Ends

Group of petroleum products with the lowest boiling temperatures, including gasolines and distillate fuels.

Lignite

A type of brown coal with low heating value primarily used for electricity generation.

Liquefaction Facilities

Facilities where natural gas is cooled down under high pressure to create LNG.

LLS

Light Louisiana Sweet. A US crude oil.

LMP

Locational Marginal Pricing. A method of pricing the cost of congestion into electricity prices. LMP aims to encourage the efficient use of the transmission system by assigning costs to users based on the way energy is delivered.

LNG

Liquefied natural gas. Natural gas converted to a liquid state by pressure and severe cooling, and then returned to a gaseous state to be used as fuel. LNG is moved in tankers, not via pipelines. LNG is predominantly methane and artificially liquefied; not to be confused with NGLs (natural gas liquids).

Load

The amount of electricity delivered or required at any specific point or points on a system. The load of an electricity system is affected by many factors and changes on a daily, seasonal, and annual basis, typically following a pattern. System load is usually measured in megawatts (MW).

Long Position

When a trader buys a commodity, in the hope that its value will go up, he/she is said to be long.

M

Margins

A deposit paid on a futures transaction. Initial margin is paid, followed by top-ups as the position develops. Margins are paid to the exchange.

Marginal Cost Pricing

A system of pricing designed to ignore all costs except those associated with producing the next increment of power generation. Sometimes referred to as incremental cost pricing.

Mark-to-Market

To revalue futures/option positions using current market prices to determine profit/loss compared to current market prices. The profit/loss can then be paid, collected, or simply tracked daily.

Market Risk
The risk that the value of an asset—commodity, energy source, or energy—decreases.

Megawatt (mw)
A unit of electrical power equal to one million watts or one thousand kilowatts.

Megawatt-Hour (MWh)
One million watt-hours of electricity. A unit of electrical energy which equals one megawatt of power used for one hour.

Mineral Rights
The ownership of the minerals beneath the earth's surface with the right to remove them. Mineral rights may be conveyed separately from surface rights.

MMbtu
One million British thermal units.

MMcf
One million cubic feet of natural gas.

Mont Belvieu
Mont Belvieu, Texas. The main propane trading hub along the United States Gulf Coast.

Moving Average
The mean of prices over a pre-defined period, for instance, the previous five days. The moving average for different time periods can be charted to generate short-and medium-term buy/sell signals.

N

Naked Option
A short option position in which the writer does not have the underlying commodity.

Naphtha
Straight-run gasoline fraction, often used as a petrochemical feedstock or blended further/mixed with other materials to make high-grade motor gasoline or jet fuel.

National Energy Board
The Canadian regulatory body which oversees interprovincial natural gas trade and pipelines. Located in Calgary, Alberta.

Natural Gas
Naturally occurring gas, predominantly methane, but usually contains some proportions of ethane, propane and butane.

NERC
North American Electric Reliability Corporation. A nonprofit corporation formed in 2006 established to develop and maintain mandatory reliability standards for the bulk electric system, with the fundamental goal of maintaining and improving the reliability of that system. NERC consists of regional reliability entities covering the interconnected power regions of the contiguous United States, Canada, and Mexico.

NGL
Natural gas liquids. Can include ethane, propane, butane, isobutane and natural gasoline/condensate. Not to be confused with LNG, liquefied natural gas. LNG is artificially liquefied methane, not the heavier fractions defined as NGLs.

NOC
National Oil Company.
NWE
Oil and petrochemicals market abbreviation for Northwest Europe.
NYH
New York Harbor—The delivery point for the NYMEX gasoline and heating oil contracts.
NYMEX
New York Mercantile Exchange. Also known in the energy industry as "the NY Merc."

O

Offer
An indication of willingness to sell a specified amount of a commodity at a specific price.
Off-Peak Demand
The time of the day when a power system would experience its lightest load.
OPEC
The Organization of the Petroleum Exporting Countries. Group of crude-producing countries which has used its collective weight of production since OPEC was founded in 1960 in an attempt to influence oil prices. Member countries include: Algeria, Indonesia, Iran, Iraq, Kuwait, Libya, Nigeria, Qatar, Saudi Arabia, UAE, Venezuela.
Open Interest
Open interest is the number of open contracts on a given future or options contract. Longs or shorts that have not been closed out are OI. Short-covering/profit-taking will tend to reduce OI.
Options
A contract under which the writer of the option gives someone the right but not the obligation to buy or sell an underlying commodity. Options can be over-the-counter or exchange-traded.
OPIS
Oil Price Information Service.
Out of the Money Option (OTM)
An option with an exercise prices lower than the cur rent market level of the underlying instrument. Such an option has no intrinsic value, but does have time value, as price changes in the underlying might bring it back into the money.
Over the Counter (OTC)
Bilateral markets in which contracts for futures, options and swaps are written on a tailor-made basis.

P

PADD
Petroleum Allocation for Defense District. A group of five geographic areas in the US used in reference to petroleum distribution.

Peak Demand

The maximum load during a specified period of time.

Peak Load

The maximum electrical load demand in a stated period of time. On a daily basis, peak loads occur at midmorning and/or in the early evening.

Peak Load Plant (Peaker)

A plant usually housing low-efficiency, quick response steam units, gas turbines, diesels, or pumped-storage hydroelectric equipment normally used during the maximum load periods. Characterized by quick start times and generally high operating costs, but low capital costs.

Peaking Capacity

Capacity of generating equipment normally reserved for operation during the hours of highest daily, weekly, or seasonal loads.

Petrochemicals

Chemicals derived from petroleum; feedstocks for the manufacture of plastics and synthetic rubber.

Physical Delivery

Generally, the satisfaction of a commodity contract by physical delivery at a specific point. In the futures market, the transfer of ownership of an underlying commodity between a buyer and seller to settle a futures contract following expiry.

Posted Price

Outright, non-market-related price requested by a seller of crude oil or products. Effectively, the list price.

Premium

The price paid by an option holder to an option grantor.

Prompt

A prompt cargo describes a cargo available for immediate lifting (one to two days).

Proved Developed Producing Reserves

Proved reserves that can be expected to be recovered from currently producing zones under the continuation of present operating methods.

Proved Reserves

Those quantities of oil and gas, which, by analysis of geoscience and engineering data, can be estimated with reasonable certainty to be economically producible—from a given date forward, from known reservoirs, and under existing conditions, operating methods, and government regulations—prior to the time at which contracts providing the right to operate expire.

Proved Undeveloped Reserves (PUD)

Proved reserves that are expected to be recovered from new wells on undrilled acreage, or from existing wells where a relatively major expenditure is required for recompletion.

Put Option

An option that gives the holder the right, but not the obligation, to sell a specified quantity of the underlying at a fixed price, on or before a specified date. The grantor of the option has the obligation to take delivery of the underlying instrument if the option is exercised.

Puts/Call Ratio
The ratio of puts to calls in an options market; often used as an indicator of investor sentiment.

R

Rack Price
The price of petroleum products at a refinery or wholesale loading rack. Rack-pricing is effectively cash and carry at the rack.

RBOB
Reformulated Blendstock for Oxygenate Blending. Blended with oxygenates to produced finished motor gasoline.

Refined Products
Refined products are crude oil based products that have been refined in a refinery. The products usually include gasoline, kerosene, distillates, liquefied petroleum gas, asphalt, lubricating oils, diesel fuels, and residual fuels.

Refinery, Oil
A plant, usually comprising distillation units and a variety of additional specialist units, for the manufacture of refined products from crude oil. *See Refined Products.*

Regasification Facility
A facility that accepts deliveries of LNG and processes it back to gaseous form for injection into the pipeline system.

Regulatory Risk
The risk associated with the potential for laws related to a given industry, country, or type of security to change and impact relevant investments. Also called political risk.

Reverse Tolling
When a gas pipeline recalls gas used for electric generation and diverts it to end-use markets when gas prices are higher than power prices.

Rho
The rate of change of the value of an option with respect to the risk-free rate of interest.

Roll Over
The transfer of a position from one futures period to another involving the purchase (sale) of the nearby month and simultaneous sale (purchase) of a further-forward month.

S

Settlement Price
A price established at the close of a trading day used to calculate the settlement of futures contracts.

Short Position
When a trader sells a commodity he/she doesn't own, with a view of buying it cheaper at a later date, he/she is said to be short.

Sour/Sweet Crude
Definitions which describe the degree of a given crude's sulfur content. Sour crudes are high in sulfur, sweet crudes are low.

Sour/Sweet Gas
Sour gas is natural gas which contains lethal hydrogen sulfide, and must be purified before being injected into a pipeline. Sweet gas is gas found in its natural state which does not need to be purified to remove sulfur-bearing compounds.

Spark Spread
The cost difference of converting natural gas into electricity. Is a measure of potential profit for generating electricity on a particular day.

Spot Market
A market where the commodity being traded is for immediate delivery.

SPR
United States Strategic Petroleum Reserve.

Spread
The difference between two prices, either across time or between commodities or instruments.

Spread-Trading
Buying one instrument/commodity and selling another, with a view to profiting from the change in the gap between the two markets.

Stop-Loss Order
Buy or sell orders put in through a broker, which are automatically triggered if the price moves above or below a certain level.

Storage, Oil
Typically onland tankage facilities for short- or long- term storage of crude oil products.

Storage, Natural Gas
Facilities used to store natural gas which has been transferred from its original location. Usually consists of natural geological reservoirs like depleted oil or gas fields, underground salt domes, aquifers, or in rare cases, abandoned mines.

Straddle (long)
The simultaneous purchase of a put and a call option with different maturities. This is a bet that volatility will increase; the rise in the value of one option will offset the non-productive premium paid by the other option.

Straddle (short)
The simultaneous sale of a put and a call with different maturities, with a view that volatility will decrease.

Strangle
Buying call and buying put with the same maturity.

Strike Price
The price at which an option holder has the right to buy or sell an underlying commodity and/or financial derivative.

Sub-bituminous Coal

A type of coal whose properties are between lignite and bituminous coal. Its primary use is power generation.

Swap

An exchange of streams of payments over time according to specified terms of the contract. Generally one party agrees to pay a fixed price in return for receiving a floating price from another party. If the floating price rises, the buyer of the swap receives a payment from the seller of the swap equal to the current market price minus the fixed price of the swap. If the floating price falls, the buyer of the swap pays the seller of the swap the floating price of the swap minus the current price of the swap.

Swaption

An Option to purchase (call swaption) or sell (put swaption) a swap at some future date.

T

Take-or-Pay

A clause in an energy (natural gas, electricity, crude oil) supply contract which provides that a minimum quantity of energy be paid for, whether or not delivery is accepted by the purchaser.

Therm

Unit of heat energy equal to 100,000 British thermal units (Btu).

Theta

The rate of change of the value of an option with respect to time.

Throughput

The volume of energy flowing through a pipeline, refinery or terminal.

Time Value

The time component in a premium for an option. Typically the time value of an option declines as it moves closer to expiry.

Tolling Arrangement

An arrangement whereby a party moves fuel to a power generator and receives kilowatt hours (kWh) in return for a pre-established fee. Under this contractual arrangement, raw materials or intermediate products stream from one company to the production facility of another company in exchange for the equivalent volume of finished products and payment of a processing fee.

Tolling Fee

A fee paid for use of electric generation assets used to convert fuel to power.

Transmission

The network of high voltage lines, transformers and switches used to move power from generators to the distribution system. Also used to interconnect different utility systems and independent power producers together into a synchronized network. Transmission is considered to end when the energy is transformed for distribution to the consumer.

Transparent Pricing

Transparent pricing relates to the access to available information on the trading of commodities and financial assets. Good transparency allows the public and other market participants to see the range of the transaction prices, the transaction volumes, and the prices and volume available for transactions.

Turnaround, Refinery

Scheduled events wherein an entire process unit is taken offline for an extended period for revamp and/or renewal.

U

Unconventional Gas

Natural gas that cannot be produced using current technologies.

Upstream

The part of the energy value chain focused on the exploration and production of crude oil and natural gas.

USGC

Market abbreviation for United States Gulf Coast.

V

VaR

Value-at-Risk. The VaR of a portfolio is the worse loss expected to be suffered over a given period of time with a given probability. The time period is known as the holding period, and the probability is known as the confidence interval. VaR is not an estimate of the worst possible loss, but the largest likely loss. For example, a firm might estimate its VaR over 10 days to be US$100 million, with a confidence interval of 95%. This would mean there is a 5% chance of a loss larger than US$100 million in the next 10 days.

Vega

The rate of change of the value of an option with respect to the volatility of the underlying instrument (sometimes referred to as kappa).

Volatility

The degree to which a particular price has fluctuated in the past or is expected to in the future.

VPP

Volumetric Production Payment. VPPs are debt-financing structures in which the Buyer advances funds to the Seller in exchange for a non-operating interest that is paid from a specific portion of production.

W

Watt

A measure of real power production or usage equal to one Joule per second. The rate of energy transfer equivalent to one ampere flowing under a pressure of one volt.

Wet Gas

Natural gas containing significant NGL components. Natural gasoline, butane, pentane and other light hydrocarbons can be removed by chilling and pressure or extraction.

WTI (West Texas Intermediate Crude Oil)
WTI crude is deemed to be traded at Cushing, Oklahoma. Traders typically refer to the NYMEX Light Sweet Crude futures contract as the WTI contract.

Sources:

1. Incisive Media. Commodity Risk Management and Trading Glossary, 2013.
2. Mercatus Energy Advisors. Energy Hedging & Risk Management Glossary, 2012.
3. North American Electric Reliability Corporation. Glossary of Terms Used in Reliability Standards, 2015.
4. PDC Energy. Glossary of Terms, 2015.
5. Standard & Poor's. Volumetric Production Payment (VPP) Transactions.
6. US Energy Information Agency. EIA.gov/tools/glossary.
7. Vivek Chandra. Glossary—NatGas.info/glossary.
8. Wiley. Foundations of Energy Risk Management, 2009.

Index

alternative energy, xvii, 1, 17
arbitrageurs, 52, 55-56, 61, 65, 145
ARGUS, 23, 27, 32, 81-82
ask, 34, 57, 66, 146
asset price, 106-107
asymmetrical, 103
at risk, *see* VaR
at-the-money, 107, 145

backward dated, 98, 132, 146, 148
bar chart, 114-116, 119-121, 125, 127
basis, 83-86, 89-90, 93, 146, 154
Basis Swaps, 83-92, 146
bearish, 15, 19-20, 65-66, 132, 146-147
bid, 28-29, 34, 46-49, 57-58, 66, 146
bilateral, 24, 34-35, 69, 83, 99, 102, 157
Black-Scholes, 103, 106, 146
Bollinger Bands, 120-121
breakout, 122, 125
brokers, 34, 55-56, 58, 61, 70, 79, 83, 85-87, 92, 160
bullish, 15, 19-20, 66, 147

call option, 67, 103-108, 110-111, 147, 149, 154, 160
candlestick chart, 113-114, 116-117, 125, 127
cash market, xviii, 23-26, 33, 36, 73-75, 81-82, 85, 147
cash price, 23, 25, 30, 72-74
CCP, *see* central counterparty

CDDs, *see* cooling degree days
ceiling price, 103-104, 108, 110, 114, 122
central counterparty, 34
channel, 117, 122
chief risk officer, 130, 134, 137
China, 5-7, 19, 21
CL, 44-46, 56, 78, 97-98
Clearport, *see* CME Group, 34, 45, 47-50, 85, 87-88, 92-93, 133
close-only chart, 116
CME Group, 34, 37, 41, 44-45, 47-49, 51, 63
coal, 1, 16-17, 25, 145-146, 155, 161
commercial, xviii, 11, 20, 38, 52, 65-71, 77-78, 85, 95-99, 111, 113, 141, 145
Commitment of Traders, 65, 67
commodities, xiii-xv, xvii, 1, 4-5, 11, 17-18, 23-28, 30, 33, 35-39, 43-45, 52-53, 55-56, 59, 61, 67-72, 75, 77, 79-81, 83, 91-92, 95-98, 100-106, 111, 113, 118, 120, 129, 133, 145-147, 149, 151-153, 155-161
consolidation, 124-126
consumption, xvi-xviii, 1, 3-4, 6, 11, 14, 17-18, 23, 150
contango, 98, 132, 146, 148
contract, xiv-xvi, 6-7, 22, 24, 31, 33-37, 39-41, 43-44, 46-53, 55-59, 61, 65-68, 70-84, 88, 91-92, 95-99, 102-108, 110-111, 115-116, 119-120, 122-123, 132, 134, 147-154, 157-159, 161, 163
contract specifications, 43, 45, 83
convergence, 68, 72, 75-76

165

cooling degree days, 3, 14, 147, 149
crack spread, 97, 149
crude oil, xiii, xv-xvi, 1-14, 18-19, 21-23, 25, 27, 30, 43-47, 50-51, 53-54, 56-58, 67, 70-71, 75, 78, 80-83, 89, 93, 95, 97, 102, 108-109, 115-116, 120, 123, 141, 145-150, 153-155, 158-163
currency, 6-7, 19
Cushing Hub, xiv-xvi, 2, 10, 22, 45-47, 78, 83, 89, 149, 163

daily settlement, 27, 30, 35, 51, 53, 59
delivery, 8, 22-24, 26-28, 35-37, 41, 45-51, 53, 61, 72, 75, 83-84, 88-91, 97, 146-149, 152-153, 157-161
demand, xix, 1-5, 9, 11-12, 14-16, 18-19, 22-25, 28, 32, 38, 59, 84, 87-89, 98, 146, 149-150, 157-158
derivative instruments, 33, 103, 148
distillates, 2, 11-12, 14, 141, 152, 159

economic data, 1
economy, 4-6, 14
EIA, *see* Energy Information Administration
electricity, 2, 13, 16, 19, 33, 43, 71, 95, 97, 101, 103, 145-146, 148-152, 154-156, 160-161
electronic trading, xiii, 25-26, 30, 36, 43, 53, 55-56, 62, 79, 83, 132
energy derivative, xviii, 33-35, 37, 40, 65-66, 69-70, 76, 79, 95, 103, 107, 133-134
Energy Information Administration, xvii, 2-3, 11-12, 14, 19-22, 30, 141-144, 151
energy trading, xiii, 24, 28, 32, 41, 59, 78-79, 93, 112, 129, 134, 138
Enron, xiii, 129, 133-134, 139
ethylene, xiii
Europe, 5-6, 151, 157
exchange-traded, 24, 33, 35, 67, 79, 112, 151, 157
execution, 21, 34, 55, 69-70, 78, 84, 131
exports, xiii, 1, 6, 9, 11-12, 14, 22, 145

final settlement, 50-51, 53, 72-73, 75, 80, 84-85, 92-93
financial, 25, 33, 35, 38-39, 43, 56, 66, 68-85, 88-89, 91-92, 103, 105, 108, 110-113, 129-131, 133-134, 137-138, 145, 149, 151-152, 161
financial derivatives, xviii, 39, 60, 65-67, 69, 76-77, 79, 91, 95, 98-99, 103-104, 110, 112, 129-130, 132-134, 137, 160
fixed price, 25-28, 70-77, 80-82, 84-85, 87, 90, 103, 110, 132, 135, 147, 158, 161
fixed-for-floating, 79-80, 84, 92
floor price, 103-104, 108, 110, 114, 122
forwards, 24-25, 35-36, 39-40, 67-68, 76, 90, 96, 104
frac spread, 97
fundamental analysis, 113
fundamental factors, 2, 113-114
futures, xiv-xvi, 5-7, 22, 24-25, 30, 32-33, 35-41, 43-49, 51-59, 61, 63, 65-68, 70-73, 75-81, 83-84, 88, 90, 93, 95-99, 102, 104, 106, 108, 110, 112, 130, 132-134, 147-155, 157-159, 163

gamma, 107, 152
gasoline, 11-12, 37, 56, 67, 70-71, 80, 95, 97-98, 141, 152, 155-157, 159, 162
generation, xvii, 1, 14, 16-17, 25, 146, 148-150, 152, 155, 159, 161
geopolitical, 1, 8, 11, 13, 18
Globex, 45-50, 55-56

HDDs, *see* heating degree days
head-and-shoulders, 124-125
heating degree days, 3, 14, 149, 153
heating oil, xvii, 1-3, 12, 21, 37, 43-44, 49, 53, 56, 59, 67, 70-71, 80, 95, 97-98, 157
hedge, xiv, 33, 67-68, 71-75, 77, 79-82, 85, 89, 95, 97, 100, 108, 139, 148, 153
hedgers, 55-56, 59, 61, 68, 76, 78, 80
Henry Hub, 48, 73-75, 80, 84-90, 92-93, 97-98, 153
high-frequency trading, 21, 65, 76

HO, *see* heating oil
hurricanes, 3, 14, 21
hydroelectric, xvii, 14, 16, 151, 158
hydrofracturing, xiii, 10, 97

ICE, *see* Intercontinental Exchange
ICE Futures-Europe, xix, 12, 36
IEA, *see* International Energy Agency
imports, 4, 11, 13, 141, 145, 154
indexes, xviii, 19, 23, 26, 28, 30, 85, 130
India, 5-6, 21
intercommodity, 97
Intercontinental Exchange, xiii, 23, 27-28, 31-32, 41, 83, 85, 92-93, 97, 133, 153
interest rate, 99, 106-107, 131-132
intermarket, 95-97, 100
International Energy Agency, 11, 19, 153
in-the-money, 107, 154
intramarket, 95-98, 100-101
intrinsic value, 107, 154, 157
inventory, 1-2, 11-12, 14-15, 98, 141
investors, xiv, 4, 6-7, 59, 65-66, 132
ISDA, 69, 80

Japan, 5-6, 21, 131

Keystone XL, 10

liquefied natural gas, xiii, 1, 14, 20, 22, 155-156, 159
liquidity, xiii, 34, 38, 41, 65-67, 69-70, 75-76, 89, 118, 133, 146
LNG, *see* liquefied natural gas
location, 26-29, 33, 36-37, 45, 52, 56, 72, 75, 80, 83-85, 87-90, 92, 95, 100-101, 146-148, 153-154, 160

margin, 34, 55, 59-61, 70, 78, 97, 99, 101, 131-132, 149, 155
mark to market, xiii, 59, 129-130, 134-138, 155

market makers, 34, 65, 78
marketplace, xiii, 2, 8, 17, 24-25, 30, 36-38, 43-45, 66, 69, 71, 85, 95, 106, 127, 151
mean, 114, 120, 125, 156
Mexico, 14, 156
momentum indicators, 114, 121, 125
moving averages, 114, 120, 125, 156

natural gas, xiii, xvi-xvii, 1-2, 10, 13, 17-20, 22-23, 25-33, 37, 43-44, 47-48, 53, 56, 59, 67, 70-75, 78, 80, 84-89, 92-93, 95-99, 101-103, 119, 143-144, 146-148, 150-156, 160, 162
New York Harbor, 157
New York Mercantile Exchange, xiv, xvi, 22, 33, 36-37, 39-40, 43-45, 50-53, 55-57, 61-62, 65, 67-68, 70-71, 73-75, 77-88, 90, 92-93, 96-97, 103, 109, 115-116, 134, 149, 151, 153, 157, 163
NG, *see* natural gas
noncommercial, 37, 52, 65, 67, 71-72, 113
nuclear power, 16-17, 20, 22
NYMEX, *see* New York Mercantile Exchange

oil export ban, 10
OPEC, 9, 19, 54, 78, 132, 157, 161
open outcry, 34, 36, 45-49, 52, 55, 63
OPIS, 23, 27, 32, 157
options, 32-33, 41, 54, 63, 78, 93, 98, 102-112, 130-131, 145-147, 149, 151, 155, 157, 159
order, 34, 55-59, 61-62, 66-67, 70, 78, 131, 160
OTC, *see* over-the-counter
out-of-the-money, 107, 157
over-the-counter, 24, 33-35, 39-41, 59, 67, 79, 134, 151

parallelism, 68, 72, 76
petroleum, xvii-xix, 8, 10-11, 13, 19, 21-22, 93, 141-142, 145-146, 154-155, 157-159
physical market, 24-25, 68, 73, 75, 81, 132, 147

pit trading, 55, 63, 70, 120
P&L, 130, 135, 137-138
Platts Gas Daily, 23, 27-28, 32
Platts Gas Market Report, 23, 27-28, 32, 85
Platt's Inside FERC, 28, 88
politics, 9
postings, xviii, 23, 26-27
power, xiii, 1, 6-8, 10, 13-14, 16-17, 25, 35, 71, 97, 146-147, 150, 152-157, 159, 161-162
premium, 28, 98, 103-106, 108, 110, 158, 160-161
price, xiii-xix, 1-9, 11-38, 41, 44-53, 56-60, 65-67, 69-81, 83-93, 95-99, 100-108, 110-111, 113-117, 119-127, 129, 131-135, 137, 145-162
price discovery, 24, 33-34, 36, 38-39, 43, 52, 56, 72
procedures, 51, 55, 130, 134, 137
production, xiii-xix, 1, 3, 6, 8-9, 11, 14-15, 18-19, 22-23, 28, 31, 44, 69-74, 78, 84, 132, 141, 148, 151, 157, 161-162
prompt month, 51, 59, 78, 98
propylene, xiii, 141
publications, xviii, 23, 25-28, 30-31, 40, 56, 72, 84-85, 88
put option, 67, 103-105, 107-108, 110-111, 150, 154, 158

RB, *see* unleaded gasoline
refinery, 8, 10-12, 19, 22, 80, 141, 145, 159, 161-162
Relative Strength Index, 121, 125
renewable energy, xvii, 1, 17
resistance, 114, 117, 121-125
reversal pattern, 124-125
rho, 107, 159
rig count, 11, 22
risk, xvii-xviii, 32, 34-36, 38, 41, 50-51, 56, 58, 65-73, 76-79, 83-85, 92-93, 95-101, 103-106, 111-112, 129-131, 133-139, 145-146, 148-149, 152-154, 156, 159
risk control, 130, 134-135, 137-139

risk control desk, 134-135
risk control policy, 134-135, 138
risk controls, 129-130, 133-134, 137
risk oversight committee, 130, 134, 136-137
Russia, 1, 8-9, 21

Saudi Arabia, xv, 1, 9, 157
settlement, 23, 26, 28-29, 34-35, 46-51, 57, 59, 74, 79, 81-86, 88, 91, 105, 112, 115, 125, 130-131, 135, 147, 149, 159
snow pack, 14
solar, xvii, 16
spark spread, 97, 160
speculative trading, xviii, 65-66, 76, 103, 110
speculators, 33, 37, 52, 56, 65-66, 68, 76, 78, 95, 113, 125
spot market, 23-24, 35-36, 41, 160
spot prices, 24
SPR, *see* Strategic Petroleum Reserve
storage, xvii, 2, 12, 14-16, 20, 22, 46, 50, 97-99, 143-144, 151, 158, 160
Strategic Petroleum Reserve, 10, 22, 141, 160
strike price, 104-108, 111, 149, 154, 160
strips, 46, 55, 58, 62, 68
summer, 1, 13-16, 32, 59, 62, 80, 108, 149
supercomputer, 21, 65-66, 76, 133
supply, xiii, 1-2, 4, 8, 10-15, 18, 22, 24-25, 28, 38, 56, 67, 69-72, 74-75, 77, 87-90, 99, 108, 146, 153, 161
support, 9, 10, 55, 114, 117, 121-125
symbols, 50, 52-53, 56

technical analysis, xviii, 2, 78, 113-114, 117, 125, 128
theta, 107, 161
time, xiii, 3, 7, 11, 14-15, 17, 20, 24-26, 28, 34-36, 38, 43, 45, 47-49, 51, 56-59, 69-70, 72, 80, 84, 95, 97-98, 100, 103-107, 110, 115, 120, 125, 135, 141, 145-147, 149-150, 153, 156-158, 160-162

tops and bottoms, 123, 125
transportation, xvii, 83-84, 89-90, 148-149, 151
trend, xviii, 5, 8, 114-116, 119-121, 124-125, 127, 146
trendlines, 114, 117-118, 122, 124-125

underlying, 8, 24, 33, 35, 37, 39-40, 61, 67, 81, 103-108, 111, 145, 147, 149-150, 152, 154, 156-160, 162
United Kingdom, 6, 21
unleaded gasoline, xvi, 12, 43-44, 48, 53, 56, 97
unrealized gains/losses, 135
US Gulf of Mexico, 3, 21
USD, xv, xvi, 6-7, 19, 131, 133

VaR, 129-130, 134-138, 162
volatility, xiv, 59, 106-107, 146, 153, 160, 162
volatilitydelta, 106-107, 149

volume, xiv, 15, 25-26, 28-29, 34, 45, 51-52, 57, 66, 69, 78, 80, 85, 89, 93, 99, 103, 110, 112, 114, 119, 121, 125, 132, 136, 161-162

WACOG, 26, 31
WASP, 26, 31
weather, 1-3, 12-15, 18, 21, 25, 59
Weekly Natural Gas Storage Report, 14, 20, 22, 143-144
Weekly Petroleum Status Report, 11, 19, 22, 141
weighted average price, 26, 28-30, 51
West Texas Intermediate WTI, 2, 7, 12-13, 22, 50-51, 56-57, 59, 66, 75, 80, 83, 89, 95, 97, 149, 163
wind, xvii, 16, 125
winter, 3, 12-16, 20, 28, 32, 59, 62, 89, 93, 98, 149

zero-cost collar, 108, 110